MAINTENANCE ENGINEERING AND MANAGEMENT

SECOND EDITION

R.C. MISHRA

Former Director
Aryavart Institute of Technology and Management
Lucknow

K. PATHAK

Professor
Department of Mining Engineering
Indian Institute of Technology Kharagpur
Kharagpur

PHI Learning Private Limited

Delhi-110092
2015

₹ 295.00

MAINTENANCE ENGINEERING AND MANAGEMENT, Second Edition
R.C. Mishra and K. Pathak

ISBN-978-81-203-4573-7

The export rights of this book are vested solely with the publisher.

Eleventh Printing (Second Edition)   ...   ...   **March, 2015**

Published by Asoke K. Ghosh, PHI Learning Private Limited, Rimjhim House, 111, Patparganj Industrial Estate, Delhi-110092 and Printed by Raj Press, New Delhi-110012.

Contents

Preface

The exploitation of knowledge and technology has resulted in the development of complex and sophisticated equipment and machinery in every field of engineering. These high investment systems, in turn, call for their effective utilization. The responsibilities of maintenance engineers have thus increased manifolds. They have to keep such machines and systems in working order to meet production targets or service requirements.

In the Second Edition of this book, it has been our endeavour to incorporate the latest techniques and methodologies used today to maintain the equipment and facilities at a minimum cost.

The topics such as research and development in maintenance, the role of overhauling in maintenance, and expert systems in maintenance have been included in the introductory chapters. Condition monitoring has its own significance in maintenance and this has been dealt with in this book as well. The other maintenance practices such as maintenance scheduling, computerized maintenance planning, reliability indices, quality and reliability, reliability improvement and testing, instrumented equipment monitoring, and so forth, which are essential for improving the life expectancy of the machinery, have also been added. To optimize the maintenance costs, a maintenance audit procedure has been incorporated. Maintenance information which is a crucial aspect in decision making is also included. The role of the lubricants is most important in maintenance function and, hence, the use of hi-tech oils and lubricants has been discussed. Pollution also affects the functions of equipment and machinery, therefore, the topic on pollution prevention has been given due importance in the book.

Safety is important in every walk of human life. Its significance in maintenance has been highlighted. The maintenance of some of the important electrical machines has been discussed along with the maintenance of mechanical equipment.

Decision making is an important function in any organization, which is also true in the case of maintenance. This vital aspect of maintenance and management has been presented in detail in this edition. Advances in maintenance have been presented in a separate chapter.

Authors would appreciate receiving suggestions from the readers to further improve the content of the book with respect to modern developments in maintenance practices.

R.C. MISHRA
K. PATHAK

Preface to the First Edition

Maintenance is one of the most indispensable jobs in any industrial organization. With the growth of industry and its modernization, new challenges are being faced by the maintenance personnel in their efforts to minimize the downtime and consequently ensure a longer trouble-free working life of numerous machinery and equipment. Such a necessity has brought about a major paradigm shift in the total approach to maintenance in the modern industry. Managerial skill must be incorporated in maintenance design and procedures. To this end, the engineering education system has included maintenance engineering as a part of its curriculum. The present text introduces the basics of this expanding subject to the undergraduate students of engineering and provides them with a broad view of maintenance that would assist in taking better managerial decisions wherever a situation demands.

There are only a few books in existence on maintenance engineering and management. The authors have attempted to present a merger of these two interlinked disciplines. The book contains twenty chapters starting with the basic concepts of maintenance engineering and management, types of maintenance systems, and evaluation of maintenance functions. The modern concept of condition-based maintenance has also been introduced. Assessment of reliability plays an important role in maintenance planning and execution. This aspect has been explained with examples. The importance of computers in maintenance has been briefly discussed. Maintenance planning and control, economic and safety aspects of maintenance as well as manpower planning for maintenance management have also been briefly presented, to give students an idea about the overall responsibilities of the maintenance function.

This book would not have taken a shape without the encouragement and cooperation of our family members. The assistance received from Dr. Binoy Kumar was one of the main factors that encouraged us in preparation of the manuscript. The authors sincerely express their gratitude and thank all those who helped in bringing out this book.

R.C. MISHRA
K. PATHAK

Chapter

1

Maintenance Concept

INTRODUCTION

The maintenance concept encompasses the areas of repair philosophies, maintenance support levels, manpower for maintenance, time required for maintenance, and the costs associated with maintenance. It serves the two purposes:

1. It establishes the basis for maintainability requirements in equipment/system design. For example, if the repair philosophy demands that no external help will be available at the organizational level of maintenance then the equipment design must incorporate some in-built tests so that the user can take care of the maintenance aspects as well.

2. It establishes the basis for the requirements for total maintenance support, which is important to complete the task. These requirements can be arrived at with the help of a proper maintenance analysis. Through proper analysis of the maintenance requirements and the anticipated frequencies of failures, the crew skill levels, the spare parts, the tools, and the facilities required can be worked out in advance.

The maintenance concept must be developed at the inception of a programme prior to the start of equipment design, in order to ensure that all the functions of design and support are integrated and geared towards meeting the overall maintenance concept.

Maintenance Concept Development

It is thus observed that the maintenance concept serves a twofold purpose: (1) formulation of maintainability requirements for the system to be developed and (2) provision of the facilities for supporting the system which has been developed. Therefore, the maintenance concept should be realistic in nature and should satisfactorily meet the needs of the system design engineers and the requirements of the logistic support planners.

During the development of maintenance concept, care must be taken to define the repair policy to meet out the maintenance requirements. Formulation of the repair policy will also help to make the necessary arrangements for the support system.

A high level of automation, deploying complex and sophisticated machines, characterizes the modern industries. The objective behind automation is to achieve higher productivity and profit in business. This has resulted in total dependency on machines and, therefore, it is essential that the equipment/system must remain in operation without any trouble. This requirement has changed the technology and the operating philosophy of the industry. A new body of knowledge is progressively coming up to assist the industry in meeting the challenges of maintaining the modern engineering systems. The measures taken by the industry to keep machines and operating systems in trouble-free condition, are collecively termed maintenance engineering.

After the equipment is designed, fabricated and installed, the maintenance department looks after the operational availability of the same. The idea of maintenance is very old and was introduced along with the inception of the machine. In the early days, a machine was used as long as it worked. When it stopped working, it was either repaired/serviced or discarded. In today's age, the high-cost sophisticated machines need to be properly maintained/serviced during their entire life cycle for maximizing their availability so that their operation and the resulting production is cost effective.

Maintenance is a function to keep the equipment/machine in a working condition by replacing/repairing some of the components of the machine. Sometimes even periodic examination is sufficient to keep equipment running. The maintenance concept is an outline plan of how the maintenance function will be performed. With this information available from the users, detailed procedures are drawn to concretize the maintenance concept. The executable procedures developed thus are collectively called the *maintenance plan*. The development of such a maintenance plan is one of the most important requirements of the maintenance programme that calls for a meaningful interaction between the user and the manufacturer. The feedback information thus received will allow the manufacturer to rearrange the design as per user's operating conditions as well as maintenance requirements.

The maintenance concept of equipment is related to the operational needs of the equipment, which may change from machine to machine or system to system. The operational needs of the equipment/system vary from intermittent use to continuous use. First of all, the operational concept is determined and then the maintenance concept is developed to support the operational concept. Finally, a maintenance plan is prepared in accordance with the concept thus developed, which can be more realistic.

For example, if the operational concept of a unit requires the use of ten trucks/dumpers at a time for a particular mission, the maintenance concept has to be designed accordingly to meet such requirements. To keep ten trucks/dumpers ready all the time, it may be necessary to procure twenty such equipment in the case of online maintenance, or fifteen such equipment in the case of remove/replace policy.

Thus, the scope of maintenance has grown to include not only system repair, but also system safety, economic viability, quality, and the most appropriate use of environment resources.

Evolution of Maintenance Concept

The traditional breakdown maintenance system can no longer maintain the economic lifeline of an industry. With the introduction of modern machinery the failure-based 'fix-it' (Figure 1.1) approach has been replaced by Condition Based Maintenance (CBM), Total Productive Maintenance (TPM), Reliability Centred Maintenance (RCM) in order to ensure a high in-service operational availability and reliability with passage of time.

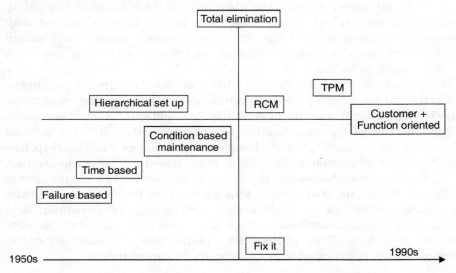

Figure 1.1 Evolution of the maintenance concept.

MAINTENANCE DEFINITION

British Standard (BS 3811–1984) has defined maintenance as follows:

The combination of all technical and associated administrative actions intended to retain an item in or restore it to a state in which it can perform its required function. This required function might be defined as stated condition.

Maintenance is the routine and recurring process of keeping a particular machine or asset in its normal operating condition so that it can deliver its expected performance or service without causing any loss of time on account of accidental damage or breakdown. In other words, maintenance means the work that is to be done, to keep the equipment/system in a running condition such that it can be utilized to its full designed capacity and efficiency for maximum amount of time.

The introduction of mechanization and automation of production systems and associated equipment, and the accompanying development of ancillary services and safety requirements, has made it mandatory for engineers to think about proper maintenance of equipment. The responsibilities of the maintenance engineers have increased considerably due to ever-growing complexity and size of industrial organizations. It has become very important to make effective use of available facilities that have been set up at high investment costs. To achieve maximum profit, it is essential to run the equipment efficiently and this is only possible when the equipment/facilities are looked after properly.

The maintenance function also involves looking after the safety aspects of certain equipment where the failure of a component may cause a major accident. For example, a poorly maintained steam boiler may cause a serious accident, and so the failure of rope of a winder used in mines.

Maintenance is also related with profitability through equipment output and its running cost. Maintenance work raises the equipment performance level and its availability but adds to its running costs. The objective of the maintenance work should be to strike a balance between the availability and the overall running costs. The responsibility of the maintenance function should, therefore, be to ensure that the production equipment/facilities are available for use for maximum time at minimum cost over a stipulated time period such that the minimum standard of performance and safety of personnel and machines are not sacrificed. These days, therefore, separate departments are formed in industrial organizations to look after the maintenance requirements of equipment and machines.

SYSTEMS APPROACH

A systems approach to maintenance involves setting up of objectives, planning, executing and controlling the maintenance functions, not as an isolated function but as a part of corporate policies and strategies aimed at achieving the well-defined overall goals of the organization. Internally, the maintenance function coordinates its activities with other functions like production and sales, and externally, it ensures availability of quality spares, trained manpower, replacement of worn-out equipment, and so forth. This approach takes into account all the relevant interacting functions, factors and facts both internal and external to the system.

Maintenance can be considered as a combination of several actions carried out in an orderly manner to repair, service or replace the components of equipment in a plant or a system, such that they continue to operate for a specified time period. In short, the function of maintenance is to maximize the availability of equipment/facility for production or operational use.

Availability

The availability of a machine or service over a specified time can be represented as

$$A = \frac{UT}{UT + DT} \tag{1.1}$$

where

A is the availability of the machine over a specified time
UT is the machine's uptime during the specified time
DT is the machine's downtime during the specified time

The uptime of a machine/service is thus the time for which it is actually available to complete the desired function. The downtime or outage of a machine is the period of time during which it is not in an acceptable working condition.

This definition of availability as stated in Eq. (1.1) considers only two states of a machine or service, namely working and failed. However, in actual practice there will be a spectrum of intermediate states as well. For practical purposes, idle (or ready) time of a machine or service as well as the total time lost in administrative or operational delays cannot be overlooked. Thus, it is more appropriate to express the availability of equipment in terms of its working effectiveness called the *operational availability* (OA) expressed as

$$OA = \frac{OT + IT}{OT + IT + AD + RT} \tag{1.2}$$

where

OT is the operating time
IT is the idle time
AD is the administrative or operational delays
RT is the repair time

Once the objectives are set, the system approach to maintenance involves the following:

(a) Identification of the needs of maintenance
(b) Analyzing the requirements for the above needs
(c) Determining the functional procedures for maintenance task selection, work planning and scheduling, work order processing, etc.
(d) Outlining a reporting and controlling procedure
(e) Development of supporting facilities and infrastructure
(f) Determining the cost accounting procedures
(g) Adopting a policy for training and quality assurance

Thus the systems approach is applied to ensure better maintenance performance for a particular situation.

CHALLENGES IN MAINTENANCE

The maintenance function of a modern industry faces a number of challenges such as:

◆ Rapid growth of technology resulting in current technology becoming obsolete.
◆ Advent of new advanced diagnostic tools, and faster repair systems, etc.
◆ Advanced store management techniques to incorporate modular technologies.
◆ Requirement of keeping both outdated and modern machines in service. For example, many industrial organizations have a combination of the old machines working on obsolete technology and the new systems utilizing the latest technology and equipment.
◆ To train and upgrade the skill of maintenance personnel.
◆ Analysis of breakdowns and parts failure for formulating corrective measures.
◆ Establishment of a separate maintenance organization.
◆ Developing maintenance schedules and repair and overhaul programmes.

The effective management of maintenance aspects under such challenging circumstances is often a difficult job. Besides the rectification of faults in the equipment, the activities of the maintenance department include:

◆ Upgradation of the existing plants and equipment and training of maintenance personnel to attain the required technical skills.
◆ Effective maintenance of the old as well as the new equipment to achieve a higher availability.
◆ Optimization of all maintenance functions including the cost.

- ◆ Incorporation of improvements to maintenance activities, especially in the areas of tribology and terrotechnology.
- ◆ Reconditioning of used/unserviceable spare parts wherever possible.
- ◆ Development of resources for the manufacturing of spare parts for the imported equipment/systems and other facilities.
- ◆ Setting up of an effective maintenance information management system (MIMS) for maintenance planning and execution.
- ◆ Effective utilization of the maintenance manpower/workforce.
- ◆ Setting up of in-house R&D cell for effecting improvements in maintenance practices.
- ◆ Utilization of computers in maintenance functions, particularly for spare parts management and fault diagnostics.

MAINTENANCE OBJECTIVES

The most important objective of the maintenance function is the maximization of availability of equipment or facilities so as to extend help in achieving the ultimate goals of the organization. Another important objective of maintenance is the establishment of safe working conditions both for operating and maintenance personnel. The numerous other objectives of maintenance are depicted pictorially in Figure 1.2.

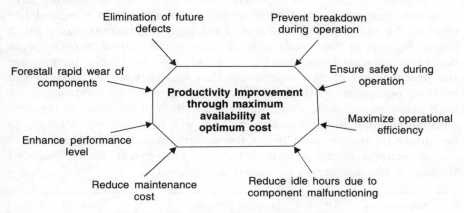

Figure 1.2 Objectives of maintenance.

In the present industrial organizations the importance of the maintenance function has increased manyfold. Therefore, its objectives should be formulated within the framework of the overall organizational setup so that the goals of the organization are fulfilled. The maintenance plans can be drawn up to provide guidelines within which the maintenance actions can be carried out effectively and judiciously without wastage of resources. For this, the maintenance personnel need to ensure that:

(a) The equipment or facilities are always kept in an optimum working condition with minimum operating costs.
(b) The time schedule of delivery to the customers is not affected because of the non-availability of machinery/service.
(c) The performance of the machinery/facility is dependable and reliable.
(d) The downtime of machinery/facility is kept to minimum in the event of any breakdown which means that the same can be repaired as quickly as possible.
(e) The maintenance cost is properly monitored to control the overhead costs.
(f) The life of the equipment is prolonged while maintaining the acceptable level of accuracy of performance to avoid unnecessary replacements.
(g) The maintenance standards in terms of quality must be set to achieve desired reliability.
(h) The maintenance records keeping and function evaluation should be taken care of.
(i) The clean and hazard free maintenance facilities are made available.
(j) The effective maintenance inventory of spare parts is maintained at each level.

One of the important objectives of the maintenance function is to keep maintenance costs under control throughout the specified life cycle of the equipment. If the cost of maintenance increases to a great extent because of non-availability of spare parts leading to consequent delays in maintenance works, it is always advisable to propose the procurement of new equipment that can be used in place of the existing equipment with minimum operating costs to compensate its high procurement cost. The important merits and demerits of the new equipment must also be analyzed. This will help the organization to formulate its replacement policy for modernization of the plant.

A general model of a maintenance system can be represented through a block diagram as shown in Figure 1.3.

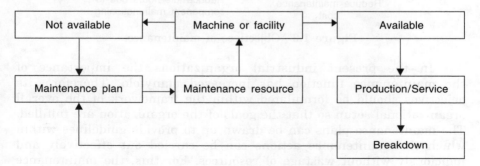

Figure 1.3 A general model of a maintenance system.

MAINTENANCE LEVELS

The maintenance of equipment/systems can be carried out at different levels depending upon the type of equipment and its place of use.

(1) **Organization level:** In the established organizations, the maintenance departments are set up to carry out the routine maintenance-related works for which the facilities must be made available.

Here normally the maintenance of minor nature such as periodic checkups, cleaning of equipment, minor adjustments, etc. is carried out. Usually, the semi-skilled maintenance personnel are associated with execution of such maintenance works of minor nature. Personnel at this level cannot repair the removed components, which are forwarded to inter-mediate level. At this level, test equipment, tools, etc. can limit the maintenance work to a lower standard as compared to the other levels.

(2) **Field level:** In this case a mobile or a semi-mobile team carries out the maintenance functions, where the equipment cannot be shifted to the workshop without its repair at the site. Here the maintenance is limited to the replacement of unserviceable assemblies. The available maintenance personnel are generally more skilled and better equipped than those at the organizational level and are charged with performing more detailed maintenance. At this level the equipment is repaired by the replacement of components/assemblies.

(3) **Maintenance workshops/depot maintenance:** This cons-titutes the highest level of maintenance and provides the support for the works, which cannot be undertaken at the other levels. All facilities related to the maintenance functions are made available at these shops to carry out major jobs including overhauls and re-building of equipment/system. In addition, the necessary standards for equipment calibration purposes are provided at this level. This level of maintenance therefore serves as a major supply base for intermediate/field level maintenance facilities.

The facilities of testing, calibration etc. are provided at these workshops. This type of maintenance system utilizes its facilities more effectively where similar type of equipment/system is used at different places.

Maintenance—Concept Optimization

Based on the system's operational requirements, alternative repair policies are decided for each equipment/system and accordingly the costs for each repair policy are estimated. From these costs the selection of repair policy with its associated cost can be done, keeping in mind that the selected policy meets the operational as well as the maintenance requirements. It is advisable to purchase new equipment/system when the operational and maintenance costs of old equipment become very high or the quality of the service is poor.

RESPONSIBILITIES OF MAINTENANCE DEPARTMENT

In consonance with the objectives of maintenance as defined above, it is necessary to lay down the responsibilities of the maintenance department. Some of the responsibilities are as follows:

Personnel Management

Proper management of maintenance personnel can improve the productivity of maintenance jobs. The job of a personnel manager involves the following:

- Maintaining availability of the working and supervisory staff and minimizing unauthorized absenteeism to avoid delays in the maintenance work.
- Arranging alternative facilities when sufficient trained manpower is not available for the jobs.
- Ensuring cordial and congenial atmosphere at the workplace, and also avoiding unnecessary hold-ups in the work.
- Maintaining discipline among the workforce so that the workers concentrate on the work allotted to them.
- Minimizing conflicts among the production and maintenance staff.

Job Distribution and Supervision

The maintenance department like any other industrial department needs to plan its operations and prepare job lists. Proper job allotment to the workers and supervisors is very important for achieving the maintenance targets. While assigning such duties, the maintenance managers need to consider the following:

- No job should be left unattended for a long period of time. The unattended jobs should be immediately allotted to the concerned persons.
- Due care should be taken during assignment of jobs keeping the time and specialization requirements in mind.
- The job instructions to the supervisors and workers should be specific, complete, and well recorded. Direct instructions

to the workers at their work place may be necessary on specific occasions. Such instructions should be given in the presence of supervisors, taking due care that no ambiguity occurs.

♦ Adequate supervision should be done from time to time to ensure the quality of maintenance work. As far as possible, supervisors should themselves get involved in the maintenance work to boost up the morale of the workers.

♦ The required spare parts and facilities must be provided well in time so that the maintenance work does not suffer.

Feedback Control

The most important function for checking the outcome of any activity is that of proper feedback. The executive of the maintenance team should get feedback on the total system under him, so that the changes, if any, can be made from time to time depending upon the operational requirements. This will help the executives in the following ways.

♦ Reinforcement of manpower in certain areas where it is needed most.

♦ Steps to be taken for procurement of special spare parts on emergency basis.

♦ To give technical guidance as and when required.

♦ To take an alternative decision when the situation demands.

Thus it is seen that timely feedback can help in proper execution of work, and breakdowns can be attended to expeditiously. Feedback helps control the downtime of equipment with minimum impact on the production programme. The following areas need proper control through feedback for successful completion of jobs.

Purchase backup: It includes timely processing of indents, timely quality control of parts; outsourcing to right sources, and implementation of effective rate contracts.

Store backup: Here attention must be focused on timely identification and issue of spares; correct codification; purchasing and receiving and inventory control.

Workshop backup: It involves manufacture of parts, overhaul of parts, quality of work and execution of modifications as required.

Operating backup: This area involves attention to standard operations, machine safety maintenance, and compliance of schedules etc.

Logistics

For efficient working of any maintenance system, provision of certain facilities such as special tools and test equipment is necessary. Such special maintenance facilities should be procured along with the equipment at the time of its purchase. As it becomes very difficult to undertake proper maintenance work for want of special tools/ equipment, these requirements must be given due attention in any maintenance setup.

Finance Management

Finance is a vital component of any system and without it no work can be done. Therefore, it is essential to budget sufficient funds for the maintenance function well in time. It has been observed that if proper care is not taken from the very beginning for careful setting up of the maintenance function, it may cost heavily at a later stage. Therefore, there is a need for proper allocation of funds for maintenance so that the requisite facilities can be provisioned well in time.

Inventory Management

During maintenance work, some spare parts/components are always required and, therefore, the required quantity of those genuine parts must be stored. Often, the non-availability of spare parts causes delays in maintenance work. Inventory management is, therefore, an important and essential duty of maintenance personnel.

TYPES OF MAINTENANCE SYSTEMS

Basically, maintenance can be divided into two groups:

    (a) Breakdown maintenance
    (b) Planned maintenance

The planned maintenance can further be subdivided into:

◆ Scheduled maintenance (SM)
◆ Preventive maintenance (PM)
◆ Corrective maintenance (CM)
◆ Condition-based maintenance (CBM)
◆ Reliability-centred maintenance (RCM)
◆ Total productive maintenance (TPM)

Figure 1.4 shows the classification of maintenance systems.

Breakdown Maintenance and Its Limitations

The basic concept of this type of maintenance is not to do anything as long as everything is going on well. Hence no work is done until a component or equipment fails or becomes inoperative. In other words,

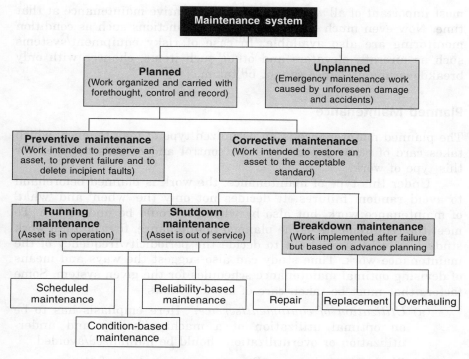

Figure 1.4 Types of maintenance systems.

the work which is required to be done in the case of an emergency failure, has to be carried out to bring back the equipment to its original working condition or even for its shipment to the main workshop for further working.

From the time machines came into existence, breakdown maintenance is still in practice and being followed in many organizations. The equipment is allowed to run till it stops working and no efforts are made in advance to prevent the failure of parts. Therefore, breakdown maintenance occurs suddenly and all the repair and maintenance work pertaining to it is only done when the equipment stops working. If this system is followed alone, it will lead to poor operational availability of the equipment as spare parts may not be readily available. Even though it may appear to be an economical proposition, work would greatly suffer if the machine is not restored to operational condition immediately. In this type of maintenance, during the repair time, no proper care is taken to know the real cause of the breakdown, which in turn may lead to frequent failures of the same kind.

With high-cost production systems being put in use, the limitations of the breakdown maintenance were realized and consequently other types of maintenance work were introduced. The most important of all was the planned preventive maintenance at that time. Now even much better maintenance functions such as condition monitoring are also available. In case of risky equipment/systems such as aircraft, elevators and other such items, chances with only breakdown maintenance cannot be taken.

Planned Maintenance

The planned maintenance is an organized type of maintenance, which takes care of other aspects such as control and records required for this type of work.

Under this type of maintenance, the work is planned beforehand to avoid random failures. It decides not only the 'when' and 'what' of maintenance work, but also by whom, it would be undertaken. To meet the requirements of the planned maintenance, first of all, a work study has to be carried out to decide the periodicity/frequency of the maintenance work. Time study can also suggest the ways and means of devising optimal maintenance schedules for the given system. Some factors that must be taken care of are as follows:

(a) *Utilization of equipment/service.* Here emphasis has to be on optimal utilization of a machine/facility and under-utilization or overutilization should be minimized/avoided.

(b) *Working conditions.* Due account is taken of the environmental factors such as humidity, temperature, and corrosiveness which may adversely affect the performance of the machine/facility.

(c) There may be some *special factors* that may affect the performance of the equipment, for example, pumping of sandy water. In centrifugal pumps, if sandy water has to be pumped, the material of the impeller should be given special consideration to obviate maintenance on account of recurrent breakdowns of an unsuitable material.

In planned maintenance, specific instructions have to be in greater detail for each type of equipment. Where safety is of paramount importance, the equipment condition should be checked everyday. Hence, the type of maintenance job will depend upon the nature of equipment and its working conditions. Planned maintenance, therefore, involves the following types of jobs that are equipment specific.

Scheduled maintenance

In this type of maintenance work, the actual maintenance programme is scheduled in consultation with the production department, so that the relevant equipment is made available for maintenance work. The frequency of such maintenance work is pre-determined from experience so as to utilize the idle time of the equipment effectively. In this way, the where and when of the maintenance work can be approximated and most efficient use of the idle time can be made. This also helps the maintenance department to use their manpower effectively. If the schedule of maintenance is known in advance, the specialists for the same can also be made available during the maintenance period. Though scheduled maintenance is costly compared to breakdown maintenance, the availability of equipment is enhanced.

Preventive maintenance

Preventive maintenance is the utilization of planned and coordinated inspections, adjustments, repairs and replacements needed in maintaining an equipment or plant. One of the main objectives of the preventive maintenance is to detect any condition that may cause machine failure before such breakdown occurs. This makes it possible to plan and schedule the maintenance work without interruption in production schedule and thus improves the availability of equipment. Under this type of maintenance, a systematic and extensive inspection of each item of equipment (or the critical parts) is made at predetermined intervals. The preventive maintenance is further divided into the following main activities.

Routine attention. This involves maintenance activities that take regular care of the machines or assets. These activities are designed to preserve the assets of the organization at a given standard of maintenance, which conforms to its financial and operating policies. Routine servicing includes cleaning, oiling, and adjusting.

Routine examination. Routine examinations are carried out to identify dormant faults or items prone to failure. This type of preventive maintenance work helps to detect faults before they can actually occur.

Preventive replacement. The preventive maintenance work comprises preventive replacement of parts and components that have a definite life. Such type of replacements help to avoid emergency situations and prolonged downtime and risk of hazards associated with sudden breakdowns.

Inspection measurements. Inspection measurements comprise jobs of preventive maintenance that aim at identifying the degradation rate of items, and such other items which are at unacceptable service conditions. This type of maintenance work requires many costly instruments and laboratory testing facilities.

Planning and implementation of a preventive maintenance system is a costly affair because during inspection all deteriorated parts/components are replaced. This type of maintenance is effectively applied in situations where the risk in operations caused by failures of equipment must be avoided. However, the higher cost of maintenance, usually gets compensated by the prolonged operational life of the equipment and reduction in operating costs. To avoid serious breakdowns, the preventive mode of maintenance is usually implemented in complex plants.

Corrective maintenance

The use of planned preventive maintenance brings out the nature of repetitive failures of a certain part of the equipment. When such repetitive type of failures are observed, corrective maintenance can be applied so that re-occurrence of such failures is avoided. These types of failures can be reported to the manufacturer to suggest modifications to the equipment.

Corrective maintenance can be defined as the maintenance carried out to restore the unserviceable equipment that has stopped working to acceptable standards. For example, an engine may be in working condition, but does not take its full load because of worn-out piston rings. Thus if the piston rings are replaced, the performance of the engine can be brought back to the specified level. The corrective maintenance, if properly carried out, will eventually bring down the maintenance costs and there will also be a reduction in downtime of the equipment.

Condition-based maintenance

This kind of maintenance is carried out in response to a significant deterioration in a unit or system as indicated by a change in a monitored parameter of the equipment or system based on its condition or performance. A condition-based maintenance policy is most suited to high capital cost equipment and complex replaceable items and also for permanent parts that deteriorate over time or due to prolong use. For this purpose, a good knowledge of failure data is necessary for effective implementation of condition-based maintenance. This process involves detection diagnosis and proper execution of maintenance planning. On indication of deterioration, the unit can be shut down at a convenient time to avoid accidental breakdowns. The following are the advantages of using the condition-based maintenance:

(a) *Safety*. Minimization of injuries and fatal accidents, to the users/operators, which can otherwise be caused by random failures of machinery.

(b) *Availability*. With increased machine availability, there will be an increase in output and also improvement in the quality of products.

(c) *Downtime*. Avoidance of unscheduled breakdowns and reduction in total downtime of equipment/system.

As condition-based maintenance is expensive, the following points need to be carefully considered before its adoption in a plant. First, not all the causes of failure can be detected in advance. Second, it involves high recurring costs because advanced monitoring techniques would be needed. Basically, there are two categories of condition monitoring:

◆ On-line monitoring which can be carried out during the operation of the machine without stopping it.
◆ Off-line monitoring which can only be carried out after shutdown of the machine.

For example, in an internal combustion engine, the temperature of the cylinders can be monitored during the operation of the engine to avoid their seizure, but the important bearings can only be inspected after the engine has been shut down.

In practice, a trade-off between the breakdown and planned maintenance costs has to be struck and the parameters influencing the policy carefully identified. Figure 1.5 shows the effects of different types of costs involved in performance of the maintenance function.

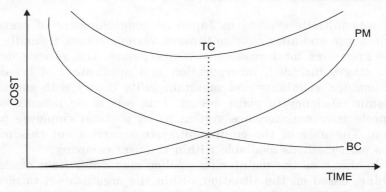

Figure 1.5   Balance of costs for maintenance. PM: planned maintenance cost, BC: breakdown maintenance cost, TC: total cost.

Reliability-centred maintenance (RCM)

It is used to identify the maintenance requirements of equipment. The RCM establishes the functional requirements and the desired performance standards of equipment and these are then related to design and inherent reliability parameters of the machine. For each function, the associated functional failure is defined, and the failure modes and effects of the functional failures are analyzed.

The consequences of each failure are established, which fall in one of the four categories: hidden, safety or environmental, operational, and non-operational. Following the RCM logic, preemptive maintenance tasks, which will prevent these consequences, are selected, provided the applicability and effectiveness criteria for preventive maintenance are satisfied.

The applicability requirements refer to the technical characteristics for possible maintenance tasks and the frequency at which these should be carried out. The effectiveness criteria depend on the consequences of the failure, probabilities of the multiple failures for hidden failure consequences, acceptable low risk of failure for safety consequences, and non-operational consequences. When the requirements for planned maintenance (PM) are not fulfilled, default tasks include failure finding (for hidden failures, possible redesign of equipment, procedures and training processes) and no-schedule maintenance.

Reliability-centred maintenance was first applied to critical, problematic areas to reduce the maintenance cost in a food processing plant. This also helped in reducing the downtime of the equipment, besides achieving reduction in scheduled maintenance load and effecting improvements in system/equipment availability.

Total Productive Maintenance (TPM)

This was initially evolved in Japan as a development of preventive maintenance and after passing through various stages, it finally, came to be known as total productive maintenance. The various development stages included, incorporation and application of breakdown maintenance, reliability and maintainability theory with concern for economic efficiency in plant design. This was made possible with a comprehensive maintenance system based on total employee participation. The user of the equipment/system carries out this maintenance with methods available within the user company.

In this case the maintenance policy may differ from machine to machine, based on the situation within the organization to meet the cost criteria. The conflicts of production and maintenance departments are minimized under this system of maintenance, as each one of them is responsible for the job. Improvements in this method are possible in order to achieve a zero equipment failure situation and consequently, maximum equipment/system availability, productivity and quality of production. The reliability of the equipment also improves under this system of maintenance. It also eliminates the dependency on outside expertise in equipment maintenance. The objective of total productive maintenance is to improve maintainability and practice preventive maintenance for the complete life cycle of the equipment/system. The benefits derived from TPM include the following:

- ◆ Reduction in unexpected equipment/system failures
- ◆ Reduction in direct maintenance costs
- ◆ Reduction in quality costs, both internal and external

- Reduction in work-in-process inventory
- Improvement in labour productivity

TPM is not only profitable but also creates a more pleasant working environment, a workplace full of vitality. The identifying characteristic of TPM is the emphasis it puts on the elimination of the 'six big losses', namely:

(i) due to breakdowns of production equipment
(ii) due to defects and defective products and components
(iii) due to set-up time
(iv) due to adjustments of equipment
(v) due to idling and minor stoppages of equipment
(vi) due to equipment operating at lower speed which, in turn, results in lower productivity and yield losses.

In essence, the TPM programme calls for the following types of activities:

- Group activities to eliminate the six major equipment related losses.
- The involvement of production operators in periodic maintenance activities.
- Restoration of all equipment to optimal operating conditions and elimination of accelerated deterioration of equipment.
- Increased efficiency and cost effectiveness of maintenance work through better planning, scheduling and control.
- Activities directed at improvement of maintainability of all existing equipment/systems.

BENEFITS OF MAINTENANCE

The high involvement of capital cost in any production system calls for proportional returns from the equipment. This is only possible when the equipment keeps delivering its normal performance. It is often noticed that the maintenance schedules provided by the manufacturer do not yield the required results in terms of the production output and the life of the equipment. Thus it becomes necessary to properly maintain the equipment with extra caution and care in order to achieve the desired levels of production or service. The following benefits can be derived from a well-organized maintenance system:

- The minimization of downtime of equipment/systems
- Improvement in total availability of the system
- Extended useful life of the equipment
- Minimization of operating costs of equipment
- Minimization of maintenance costs in the future
- Safety of the personnel
- Improved confidence of the operating personnel/users.

The consequences of downtime can be very serious when the machine is working in a production line, as its failure will shut down the total system. Following a proper maintenance schedule with adequate back-up supply of required spare parts can drastically reduce the downtime and thus improve the equipment availability. The normal wear and tear of equipment can also be reduced by proper maintenance. In certain cases, the safety of the personnel is of prime importance and this also can be assured by proper planned preventive maintenance. For example, all the aircraft systems need to be inspected before and after a flight as safety of the passengers as well as that of aircraft is of prime importance.

EFFECTS OF MAINTENANCE

Maintenance, being an important function in any production system, has far-reaching effects on the system. If the right kind of maintenance function is not selected for a particular environment, it may lead to the serious problem of either over-maintenance or under-maintenance. The selection of a particular maintenance policy is also governed by the past history of the equipment. Cost effective maintenance will help boost the productivity in a production system. It is, therefore, most important for the team, involved in maintenance work, to know how much to maintain.

The life of any equipment also depends on the nature of maintenance function. It is known from experience that optimum maintenance will prolong the life of the equipment, and on the other hand, carelessness in maintenance would lead to an early failure as well as reduced life of the equipment. Further, proper maintenance will help to achieve the production targets. If the availability of the equipment is high, the reliability of the production system will also be high.

Another important effect of the maintenance function is the working environment. If the equipment is in good working condition, the operator feels comfortable to use it, otherwise there is a tendency to let the equipment deteriorate further, and the relationship between the maintenance group and the operating group is strained. To get the desired results in a maintenance operation, there should be selective development of skilled, semi-skilled and unskilled labour and proper division of responsibilities among them in order to make full use of the skilled workforce available. Evaluating maintenance is not easy, it is therefore, important to ensure that the work being done is, in fact necessary. The schedules and programmes must be justified and based on the practical experience under the similar conditions.

CONCEPT OF MAINTAINABILITY

Maintainability is defined as the probability of restoration of a failed device or equipment or asset to operational effectiveness within a specified period of time through the prescribed maintenance operations. It is associated with the design of the assets to be maintained and is the measure of the ease of maintenance. The parameter for expressing the maintainability is the *Mean Time to Repair* (MTTR), i.e. how much time is required to restore the equipment/system to its working condition..

The concept of maintainability is different from reliability. Reliability is defined as the probability that an asset or a system will operate satisfactorily for a pre-determined period of time, under the working conditions for which it was designed. The parameters expressing reliability are *Failure Rate* (FR) or *Mean Time Between Failures* (MTBF).

Though technological developments have contributed to increased reliability (accuracy, sustained rate of operation, etc.), the corresponding complexity of the parts has made the maintenance job rather difficult. Today, it is usually not possible to repair the parts of equipment and therefore they have to be replaced. The failure reports from the users are made available to the manufacturers to introduce modifications, which make the equipment more reliable, though at the same time more complex and difficult to maintain.

The present-day designs emphasize reduction in the number of parts in order to improve reliability and leave room for easy maintenance. But this is not possible in the case of non-maintainable products, where no design provision is made for servicing or repair and all defective units have to be replaced.

Sophisticated equipment are good as long as there are no failures, but when failures occur, though occasionally, due to complex nature of equipment, the same may have to be dismantled fully to detect the defective parts. This requires the availability of specialized equipment and trained workers. The total time and cost required to service/repair such equipment are very high and the benefits of the reliability are reduced. The life cycle cost of maintaining complex equipment is high and in certain cases it equals the cost of the equipment. In the defence sector, about 25% of its budget is spent on maintenance, looking into the importance and the requirements of the equipment/system. In the developed countries the concept of maintainability came into existence in the late 1960s. Where the cost of maintaining the state-of-the-art equipment exceeded the targets, the concept of maintainability was introduced. Therefore, maintainability can be defined as the characteristic of equipment design and installation, which is expressed in terms of ease and economy of maintenance, availability of equipment, safety, and accuracy of performance parameters of equipment. Its aim is to design and develop a system or equipment that can be easily maintained at a reasonable cost with minimum resources, without affecting the performance and safety of the equipment.

The concept of maintainability was thus evolved to enable designers to provide effective support to maintain the equipment. This idea was thought of because of the complexity of design, the size and number of parts that make a system and, therefore, the maintainability features form an important part of the system design today.

The limitations of the maintainability are that, firstly, it cannot be built in complex machine systems because of design considerations and, secondly, it does not improve the performance of the equipment.

PRINCIPLES OF MAINTENANCE

Effectiveness of maintenance depends on systematic conductance of various interrelated activities. To achieve the overall objectives of a maintenance system, certain established principles must be followed. These principles help guide the maintenance staff in the efficient discharge of their duties. The main areas of work governed by these principles are discussed below:

Plant Management in Maintenance Work

The main role of the maintenance function is to provide safe and effective operation of production equipment in order to achieve the desired production targets on time economically. Sometimes it is observed that until and unless there is breakdown of equipment, the production department does not follow the maintenance programme because of its keenness to accomplish the job in stipulated time. Therefore, it is necessary that all the departments cooperate with each other to meet the organizational goals.

Maintenance Objectives versus Plant Production

The plant production department is normally assigned production targets, in the light of which the maintenance team must formulate its assigned objectives to match the total objectives of the organization. It is observed that, in practice, the production department tends to overlook the maintenance functions in its eagerness to meet the production targets. Such an attitude may lead to serious circumstances, resulting in equipment failures or accidents. The programme of scheduled maintenance must be followed regularly in order to reduce the number of breakdowns during the operating life cycle of the equipment.

Establishment of Work-order and Recording System

The preparation of the work-order for the maintenance function is an important task, since it indicates the nature of the work to be performed and the series of operations to be followed for a particular job. It is essential that at every workstation, proper records/entries be maintained to monitor the health of the equipment as well as its lifespan. Presently, in many organizations this work is being performed with the help of computers.

Information Based Decision Making

For successful completion of any work, the importance of a reliable information system cannot be overlooked. Such a system may be used for decision-making in respect of manpower and spare parts requirements. Therefore, data entries for any type of breakdown must be properly recorded in time, so that the same may be utilized as and when required.

Adherence to Planned Maintenance Systems

It is generally observed that the instructions of the manufacturers are not given due consideration while the equipment/services are in use. This often gives rise to unscheduled breakdowns of equipment. If the manufacturer's instructions are rigidly adhered to, it will not only enhance the life of equipment but also ensure proper use of manpower.

Planning of Maintenance Functions

Though minor jobs related to maintenance can be done without any prior planning, major jobs must be planned in advance for effective utilization of manpower/resources and to obviate incurring of heavy shutdown costs because of equipment breakdowns. After completion of each major maintenance routine, the cost incurred and the performance obtained must be compared with standard figures for justification.

Manpower for Maintenance

All jobs involved in maintenance functions must be carefully examined through a time and motion study for the calculation of manpower requirements. This can be done in the case of overhauls, major component replacements, emergency and unscheduled repairs. The man-hours calculated for all such jobs can be made use of, for working out the size and composition of the workforce. Initially there may be some problems, but with experience these calculations can be made fairly accurately.

Work Force Control

As the nature of maintenance work is quite unpredictable, the cost of maintenance function can be kept under control only with proper monitoring of the work force. For this purpose, sufficient field data are required to be collected regularly, accurately and completely. These data can be made use of for work force control.

Quality and Availability of Spare Parts

At the time of purchase of an equipment the manufacturers do supply some spare parts, but these are usually not sufficient for all types of faults encountered in practice and for the whole life of the equipment. Therefore, it is essential to stock a complete range and scale of genuine spare parts to sustain the functionalities of the equipment without endangering safety.

Training of the Maintenance Workforce

In the case of modern sophisticated equipment, the nature of the maintenance function also needs to undergo changes. Whenever new equipment is purchased, the maintenance personnel along with the operating personnel must be trained. The supervisory staff also requires appropriate training to assure quality of maintenance. On-job training of new recruits under skilled workers and supervisors is practised in many industries. Evaluation of training is also essential to meet the growing challenges of maintaining sophisticated and delicate equipment.

RESEARCH AND DEVELOPMENT IN MAINTENANCE

Due to advancements in technology, the concept of maintenance to improve and to enhance the life of the equipment/system is growing fast. Lots of studies have been carried out in the developed countries to improve the reliability of the product, and the concept of maintainability is also gaining significance to make the maintenance job easy and time saving. But the following areas need proper attention of all concerned to bring out substantial improvements in working environment.

1. **The influence of maintenance on design and selection of plants:** The necessity and importance of maintenance

can be decided by the designer, manufacturer and users in consultation with each other in order to improve the availability and reliability of equipment/system. The sole aim here should be to keep the operating and maintenance costs to minimum.

2. **Use of available information:** The information available on maintainability, reliability etc. can be used to improve the performance of the equipment/systems.

3. **Training requirement:** Time-to-time training must be imparted to the practising engineers and technicians to make full use of the advancements in technology.

4. **Standardization of product:** Efforts should be made for the standardization of parts/components particularly when they are imported, so that the users face minimum problems. Interchangeability of parts must be given due attention to make its use feasible.

5. **Spare parts management:** The availability of genuine spare parts is the main concern of the maintenance engineers and the determination of its quantity is a difficult task. Therefore, efforts should be made to use the facilities of experts in the field.

6. **Development of maintenance facilities:** The advancements in technology have made the task of the maintenance engineers more challenging because of the size and the volume of the equipment/system. This calls for timely maintenance to keep the downtime to as minimum as possible. For effective working of maintenance groups, special facilities required must be developed to face the challenges in the field of maintenance.

7. **The concept of condition based maintenance:** Attempts should be made to use the modern techniques in the field of maintenance in the case of costly and sophisticated equipment/systems. The concept of condition-based maintenance is quite new and helps in maintaining machinery effectively. Though the initial cost of this system is more but better availability and safety of the equipment is assured.

8. **Concept of contract maintenance:** It is noticed that the nature of the maintenance function is complex in nature and therefore needs services of trained and qualified persons. With regard to specialized equipment/systems the contract maintenance works well since qualified and trained personnel are engaged in the job. The example of this is the computer industry.

9. **Development of indigenous spare parts:** In case of imported equipment/systems it is difficult to procure and preserve adequate quantity of spare parts and therefore efforts should be made to develop material and spare parts at the organizational level. Sometimes the manufacturers stop production of a particular type of equipment the same equipment is in use. Obsolescence is another factor, which calls for the development of technology to meet out the challenges.

ROLE OF OVERHAULING IN MAINTENANCE

The equipment/systems used in industry can be put under two groups: (i) equipment with diminishing efficiency and (ii) equipment with constant efficiency. The first category of items deteriorate with time, resulting in increase in operating cost whereas the second category of items operate with constant efficiency for a certain period of time and fail suddenly.

When the concept of overhaul is applied to the equipment with diminishing efficiency the operating costs are brought down after every overhaul and the period of useful life is extended. The period of overhaul must be decided to actually derive the gain of the above concept.

EXPERT SYSTEM IN MAINTENANCE

It is evident that the cost of computer-assisted systems is very high and therefore, enough care should be taken to ensure their smooth operation for the maximum possible period of time. The unexpected equipment failures entail significant financial losses in manufacturing systems that are operated on highly optimized schedules. In order to avoid uscheduled failures, the concept of preventive maintenance with expert system can be used.

An expert system is a program, which has a wide base of knowledge in a restricted domain, and uses complex inferential reasoning to perform tasks, which human experts could do. The important parts of the system are:

Knowledge base: It encompasses knowledge about the domain problem. Knowledge representing methods are: production rules, frames, semantic nets and predicate logic.

Inference engine: The inference engine is the problem-solving control strategy or rule interpreter, which applies the knowledge in the knowledge base of the problem. The strategies are forward and backward chaining.

Two fundamental modules support the system construction. They are explanation system and knowledge based management system.

Three interfaces complete the architecture of an expert system. They are user interface, knowledge engineer interface and external interface.

Development of expert system

The system can be developed in high level languages like prolog, LISP, etc. But here, one has to put in a lot of efforts in developing the program to make it user friendly. It is very easy to develop an expert system by using the expert system shell available in the market.

Validation of system: The system is evaluated by the empirical approach. The approach is to test a representative set of selected problems. The explanation system facilitates easy verification of the correctness of the expert system based on a comparison with his own decisions. This expert system provides guidelines for the maintenance staff. The periodic/preventive maintenance that should be provided for the numerically-controlled machines, has been kept in the knowledge base of an expert system shell. The expert system gives some recommendations as per the user's input and also comments on the maintenance.

QUESTIONS

1. Define maintenance and illustrate its systems approach.
2. What should be the objectives of maintenance management for successful working of a maintenance department?
3. Briefly enumerate the challenges of maintenance function.
4. Discuss the responsibilities of the maintenance department in a well-established organization.
5. Explain the duties and responsibilities of a maintenance engineer.
6. Describe the types of maintenance being practised in the present-day industrial setups.
7. Illustrate the principles of maintenance for improving the equipment/system availability.
8. What are the consequences of insufficient maintenance? Explain.
9. Write the disadvantage of excessive maintenance.
10. Explain the importance of research and development in the field of maintenance.
11. Discuss how an expert system can improve the maintenance in case of NC machines.

Chapter

2

Planned Preventive Maintenance

INTRODUCTION

The Planned Preventive Maintenance (PPM) encompasses a variety of work schedules ranging from simple visual examinations to internal inspection of machinery. It also includes the time-based maintenance or the cyclic changes of plant irrespective of its operating conditions. The engineering management of any industry relies on the judgment of its engineers to plan remedial work in advance. The actions that are initiated and carried out before the occurrence of failures are categorized as *preventive maintenance*. These actions are deterministic in nature and need to be carried out at fixed intervals. They can be scheduled and carried out under the preventive maintenance programme. The planned preventive maintenance system is basically the utilization of planned and coordinated inspections, adjustments, repairs and replacements made in maintaining an industrial plant. During PPM, the prescribed standard operations of lubrication, repair, adjustments, etc. are carried out with a view to eliminating faults or alarming conditions of machines or any other asset while the defects are still in a nascent stage. The purpose of PPM is to implement the necessary and timely servicing with a view to preventing the unscheduled interruptions or the undue distortions of an operating equipment, building, or asset.

The objective of the Planned Preventive Maintenance (PPM) is to carry out the necessary and timely servicings in order to prevent unscheduled breakdowns or undue deterioration of equipment or facilities. As the work of preventive maintenance is well coordinated, the downtime of equipment/machines due to failures can be minimized and thus untimely expensive repairs eliminated. The downtimes of high-cost capital equipment are also very costly; it is, therefore, preferable to carry out maintenance on a schedule basis to avoid breakdowns. Such a system helps in achieving optimum utilization of available manpower, machine time and in ensuring prolonged life of equipment. However, there should be a proper justification for having a planned maintenance programme for any equipment, since the cost of this maintenance is usually very high even though the interruptions in the operating cycle of the equipment can be minimized. The cost effectiveness of such a programme can be achieved by concentrating on that equipment which is considered to be critical. Units whose failure will make the equipment inoperative are the critical units. Units which are procured through high capital investment, are also considered critical.

With the above definition of critical units, attention can be focused on such units for proper maintenance, so that benefits in terms of reduced downtime and high production are achieved. On the other hand, the units that are not classified as critical are subjected to random routine maintenance only. The basic idea is to concentrate on selective units to optimize the maintenance programme.

It is customary for the manufacturers to prescribe a maintenance programme for their equipment. However, the same may not be satisfactory in a specific site condition or operating conditions and may thus require some modifications. Such modifications and alterations are possible to incorporate through the planned maintenance system. The necessity to have a preventive maintenance programme is, therefore, due to economic considerations, attributable to high cost of capital investment and the returns expected from it. The useful life of equipment can be extended through proper maintenance as well as its availability maximized.

Although profit maximization is most desirable, it mostly depends upon the operating conditions and requirements of the equipment used. The most appropriate objective to adopt would be maintenance cost minimization for indirect maximization of profits.

The cost of repairs can be reduced either by reducing the frequency of failures or by reducing the cost of each maintenance action, or by reducing the both. It may be possible to reduce one but by only increasing the other and still achieve some reduction in total cost.

Some guidelines to maintain the equipment can be formulated and a time schedule for checking the equipment prepared. There are a number of maintenance policies that can be used individually or in combination for each unit of plant/facility. Such maintenance policies can be classified as:

◆ Fixed-time maintenance
◆ Condition-based maintenance
◆ Operate-to-failure
◆ Design-out maintenance
◆ Opportunity maintenance

The fixed-time maintenance means maintaining an asset after a defnite time interval. In condition-based maintenance, the status of an asset or its components is monitored. If the condition of such monitored item deteriorates beyond a certain limit, maintenance actions are initiated. Operate-to-failure policy does not follow any preventive maintenance. In this scheme, the machine is used until it fails. In the case of design-out maintenance, the possible causes of failure are considered at the time of designing the component. Machines so designed do not have frequent maintenance problems. In case of opportunity maintenance if the equipment fails for some reason or the other, the components/parts which are critical are also examined to avoid failures in the future.

SCOPE OF PREVENTIVE MAINTENANCE

The purpose of preventive maintenance is to establish:

- ◆ Routine maintenance of jobs, through visual inspection and testing
- ◆ Standard tasks covering routine work
- ◆ Inspection and overhaul schedules for all major equipment
- ◆ Spare parts requirements
- ◆ Manpower assignment as per requirements

If the above points are well taken care of, it will enhance the system/equipment availability and therefore affect a reduction in downtime. The preventive maintenance is carried out in two ways. In the first method, the maintenance is implemented with the unit remaining in operation. And in the second method, the unit is stopped for the maintenance work. The preventive maintenance works are deterministic and so they can be scheduled and usually carried out separately according to a preventive maintenance checklist. In such a checklist the frequencies of inspection of various components of the unit are provided. These components are therefore inspected as per their prescribed interval and their condition recorded. Maintenance experts need to identify the components and the condition indicators for better results from planned preventive maintenance. Table 2.1 shows, for example, a few mechanical systems with the indicators to be inspected during the preventive maintenance programme.

Table 2.1 Mechanical systems and indicators to be inspected during planned preventive maintenance

System	Indicators
General monitoring	Vibration, tightness, alignment
Power system	Fuel level, performance measurements, cleanliness
Cooling system	Temperature, leakage, filter condition
Lubrication system	Temperature, pressure, flow rate, lubricant properties, fuel consumption rate
Transmission system	Depends on the type of the system
Joints, fixtures, connections	Operational effectiveness
Engine	Power supplied, sound level, temperature, fuel consumption rate

ELEMENTS OF PPM

To plan for preventive maintenance, the following characteristics must be known for each component/system. These can be considered as elements of PPM.

Deterioration Probability

If a component/part deteriorates frequently, its performance decreases until it fails during operation. Such a part may be inspected at scheduled intervals of time and if found deteriorated beyond its limits, it may be replaced. The probability that at a given inspection the component/part would need repair or replacement is termed *deterioration probability*. This probability depends upon factors such as physical standards, techniques used for inspection and on the time and usage of the component/part between inspections. During scheduled inspections the causes of deterioration can also be established through analysis, which will help to avoid or minimize their occurrences during the further use of the equipment.

Failure Probability

The probability that a component/part will fail in a given period is called its *failure probability*. The expected number of times that a failure would occur between preventive inspections of the component is the sum of the probabilities from each of the time periods that occur between inspections. This probability and the deterioration probability are derived from transition probabilities that represent relative frequency, a component change from one state to another, during the given time period. The failure probability can be determined by using the Markov chain processes.

Example: Suppose in a particular plant 15 per cent of equipment is in failed state and the remainder in the repaired state. Data from the previous records indicate that 10 per cent of the failed equipment could not be repaired and 5 per cent of the repaired equipment failed. Assuming this as the constant maintenance performance, determine the percentage of equipment that would be in the failed state in the second year.

Solution: Using the Markov Chain, the transition matrix

$$p = \begin{bmatrix} 0.10 & 0.90 \\ 0.05 & 0.95 \end{bmatrix}$$

and the state of the system in the first year is

$$S_1 = \begin{bmatrix} 0.15 & 0.85 \end{bmatrix}$$

Therefore, in the second year, $S_2 = S_1 p$

$$= \begin{bmatrix} 0.15 & 0.85 \end{bmatrix} \begin{bmatrix} 0.10 & 0.90 \\ 0.05 & 0.95 \end{bmatrix}$$

$$= \begin{bmatrix} 0.0575 & 0.9425 \end{bmatrix}$$

Thus, the percentage of equipment that will be in the failed state will decrease to 0.0575, i.e. 5.75 per cent.

During preventive maintenance schedules, a number of components/parts are regularly checked and inspected to assess their actual condition. If the number of such items increases, it will increase the maintenance workload and the manhours accordingly. This should be given due care during the formulation of the planned preventive maintenance schedules so that the idle time of the equipment/system is minimized.

During an operation, whenever a component fails, the system will become inoperative necessitating immediate repairs. Depending upon the prevailing conditions, the emergency repairs are undertaken to make the system operational. It is also possible that because of partial failure of a component, the same can be used at reduced efficiency till the repair work is undertaken but this should not be practised as a regular feature of the maintenance system since it can affect the service life of the system.

Cost of Periodic Maintenance

If the equipment used has a high capital cost, and its usage is regular in nature then the major cost associated with the maintenance system is the downtime cost for the want of regular check up. The other costs, such as the labour and material costs required for a periodic maintenance setup are negli-gible compared to the downtime cost. The setup cost is incurred each time the periodic maintenance is undertaken. During the maintenance planning stage, as far as possible, the idle time of the equipment must be utilized to carry out the periodic checks in order to keep the total downtime of the system to a minimum level.

Expenditure towards maintenance personnel

A maintenance crew is kept to maintain the systems in their operational condition. Such personnel are required to perform all related works such as preventive maintenance, inspection, repair, diagnosis of emergency breakdowns, and numerous routine tasks including the maintenance of tools required for the maintenance work. The cost involved in this work is fixed but will depend upon the number of crew members and their skill levels. To minimize this cost, a sound planning is essential and therefore care should be taken to select the persons of requisite specializations in order to keep the overall maintenance costs at a minimum level.

Cost of system downtime

The downtime cost indicates the production losses due to outage of the system for maintenance/repair and will largely depend upon the type of equipment used and the type of maintenance system followed. As mentioned earlier, this cost will be very high in the case of capital-intensive equipment. Due cognizance of downtime cost should be taken while deciding a maintenance policy.

Practice of good PPM results in the reduction of downtime of equipment/system. Downtime cost is calculated taking into account the loss of production, the expenses on idle manpower, the loss of goodwill, and the cost of restoration, etc. Suppose in a particular situation, PPM can reduce the unscheduled stoppages to 30 minutes per week. Let us assume that the unscheduled stoppages earlier used to be two hours per week. The downtime cost reduction can thus be calculated as follows:

Reduction in downtime cost = [Unscheduled downtime (h) per week
− reduction in downtime (h)]
× downtime cost per hour.

Consider ₹ 10,000 per hour as the downtime cost for the above assumption.

Therefore, the reduction in downtime cost

$$= ₹ (2 - 0.50) \times 10,000 \times 52 \text{ per year}$$
$$= 1.5 \times 52,000 \text{ per year}$$
$$= ₹ 780,000 \text{ per year}$$

Supporting Services

For effective planned/preventive maintenance, it is necessary to cater for adequate supporting services which will be of paramount importance for the success or failure of any policy and therefore should be given due cognizance. Some of such facilities are mentioned below:

Facility register

Each organization should maintain a record indicating all the machines/systems held in the organization, which need to be maintained. In a large organization, ways and means should be introduced to subdivide the entire information into the sections for easy availability to the users. This can be done depending upon the maintenance needs or usage, as the needs of each section would be different.

Another classification of machines/systems can be done based on technical disciplines such as electronic, electrical, instrumentation, mechanical, hydraulic, civil, etc. In a similar way, the maintenance crew can be identified for each group. Nowadays computers are used to store all the relevant information regarding the performance and failure of machinery, which can assist in decision making as and when required for effective maintenance of the same.

Equipment record sheet/card

This record is very useful when the maintenance work is of repetitive nature as it contains all the relevant information such as the details of equipment in respect of services needed, spares supplied, lubrication used and other specifications. The information given on the record sheet/card proves very useful for maintenance work planning and scheduling. From this record, information can be transferred onto the equipment history card known as the *log card*. The log card should move along with the equipment, whenever the equipment is shifted from one place to another for any work. Such cards are required to point out the equipment details, which can be used by the production as well as the maintenance personnel. There are many types of history cards used in practice, however, one type for an electrical equipment is shown below.

ELECTRIC MOTOR

Equipment code:

Code number:

Plant identification number:

Location of the department/office:

Date of commissioning:

Date	Details of parts replaced	Manhours spent	Downtime	Signature

Maintenance schedules

These schedule indicate the nature of work to be done, at what time and the estimated time required to complete the job. Separate schedules are prepared for each job, and carried out on each component/part as per the facility register. For preparation of schedules, detailed information is required which can be taken from the records. One schedule can be sufficient for small components/ parts, whereas for large-size equipment separate schedules are prepared for each of its component/parts such as engine, motor, hydraulic system, and controls, etc. The schedules should contain full information regarding the work to be done. The coordination of associated group is important for successful completion of maintenance jobs where separate schedules are drawn for different components of the same equipment.

Work specifications

The exact maintenance work to be performed varies according to the type of equipment and the type of maintenance selected. The detailed work specification to be performed is specified on the job card for each of its elemental activities. The job card gives the estimated time for each item of work to help in manpower planning and control. Due care should be taken while registering the activities on the job card so that the maintenance personnel have clear understanding of the jobs which need their attention.

Maintenance records and documentation

The maintenance records are necessary tools for better planning and they aid the decision-making process as well as improve efficiency of the system. It is necessary to record the relevant maintenance-related parameters effectively and efficiently, though this is a costly function. The maintenance information can be recorded in a variety of different formats, depending upon the depth and details to which the management decides to record such information. This information can be used during the preparation of maintenance policies and schedules, and if done properly, it will reduce the maintenance costs to a greater extent.

Spare parts inventory

The most important supporting service required for the maintenance function is the availability and issue of spare parts. At times it is noticed that the equipment cannot be repaired for want of spare parts, particularly when the equipment used has been imported and its parts/components are not available in the local market. Therefore, it is most desirable to maintain an adequate stock of spare parts depending upon the need, so that the maintenance work does not suffer. For this purpose, a proper computerized information system related to spare parts inventory is necessary to be maintained. The effective management of spare parts will minimize the downtime of equipment and ensure optimum utilization of capital investment on spare parts inventory.

IMPLEMENTATION OF PPM

The most important function of any project is the implementation and this is true for maintenance work as well. Since the maintenance function has fluctuating, multitude workload, the task of its management is really difficult. Besides the normal maintenance work, the maintenance organization may have to incorporate modifications from time to time depending upon the influencing factors, as breakdowns do occur randomly. During implementation of PPM, the following factors should be taken into consideration.

Maintenance Resource Structure

As the maintenance organization can only have a finite form, the resources for such a function will have to be balanced with the workload requirements to achieve cost effectiveness. The workload falls into two categories.

Deterministic load

This is used for planned and scheduled maintenance work and is usually applicable for a long period. It is possible to calculate the deterministic load in advance when the work contents are known.

Probabilistic load

It is used for the short-term requirements, usually for emergency maintenance work. Since its nature is of fluctuating type, it is difficult to calculate it in advance. The cost involved is more compared to that for deterministic load.

As classified above, the workload must be divided into trades, for example, fitting, instrumentation, electrical, etc. The skill levels can be improved by conducting training programmes. The proper control of workforce will reduce both indirect and direct costs for planned as well as emergency maintenance work. If the workload is mainly deterministic, it is not very difficult to calculate the size of the manpower requirements. However in a productive system, the workload usually contains large probabilistic components, which occur at random. A proper workforce size and structure will minimize the total maintenance cost.

ADMINISTRATIVE STRUCTURE

There is no ideal structure available for the maintenance work. Since the maintenance work was not given its due importance in the past, today, maintenance is considered really important due to high capital cost of equipment involved. Any structure adopted for the maintenance function should suit the complexity of maintenance work and the resources available thereof. Experience shows that the best maintenance administrative structure is the one, which is based on division of work into specializations. A typical maintenance administrative structure is shown in Figure 2.1.

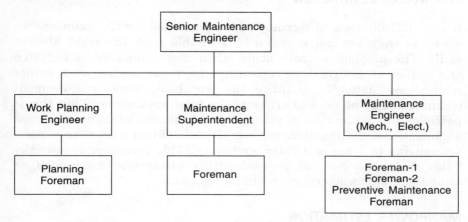

Figure 2.1 Maintenance administrative structure.

WORK PLANNING AND SCHEDULING

After the preventive maintenance schedules have been prepared, their planning and scheduling is important to meet the maintenance load. *In simple terms the function of such a procedure (planning and scheduling) is to get the right resources at the right place to do the right job in the right way and at the right time but with minimum cost.* The deterministic load consists of the preventive maintenance work and can be planned in advance, whereas the probabilistic load arises due to random breakdowns and needs proper attention for the completion of the job in time.

Job Priority

Even in planned preventive maintenance work, some methodology of prioritization must be established to reduce the workload of the maintenance department. In most of the cases, the preventive maintenance requirements are designed under three levels of priority. These are:

- ◆ Emergency operations
- ◆ Machine running operations
- ◆ Normal operations

Experience has shown that the above priorities have worked in most instances. Emergency maintenance gets top priority in most of the organizations. However, prioritization also depends upon the type of equipment used or service provided. A priority index can be drawn up for the maintenance work depending upon its importance.

WORKLOAD ESTIMATION

It is a difficult task to recruit trained personnel for the maintenance work, as they are not necessarily available with the right kind of skills. The problem is more acute when new equipment is inducted and sufficient information regarding the maintenance work is not readily available. To quantify the workload, work measurement techniques may be used to arrive at the total work content, but this is possible only in the case of planned preventive maintenance. In some organizations, the maintenance workload is subcontracted to experts specializing in some particular equipment. This procedure is specially followed in the case of general utility electronic items such as computers, photocopiers, air-conditioners, etc.

MANPOWER ESTIMATION

The maintenance organization has to undertake both planned, unplanned and emergency maintenance work. The problem here is of estimating the maintenance manpower. Knowing the total number of equipment due for various maintenance functions, a scheme for determining the manpower is formulated for this purpose. In practice, however, any planned schedule undergoes a change because of various constraints that cannot be anticipated at the planning stage. Under such circumstances it is possible to provide some flexibility in schedule, which will depend upon the past experience. It is observed from experience that any equipment coming for maintenance work will require at least two mechanics at a time and this should be considered during calculations of the manpower requirements. The backlog in maintenance work should also be taken into account during the calculation of manpower, as some of the work may not have been attended to in time due to some valid reasons such as absenteeism of workmen, non-availability of spare parts, or emergency breakdowns, etc.

SCHEDULING PPM

To derive the benefits of preventive maintenance, it is necessary to prepare maintenance schedules for all the equipment in a shop. As the workforce and the number of equipment to be serviced increase, the work of scheduling, rescheduling and that of backlog becomes increasingly difficult to plan, though the computer can be used to generate the schedules.

Once the process of schedule generation is complete, all assigned work should progress smoothly until something comes up that necessitates alterations in the predefined process. Sometimes due to budgetary constraints, rescheduling of maintenance jobs becomes necessary. It is then necessary to work out the schedules in such a way that backlogging is avoided. This is mostly necessitated when a fixed scheduling process is followed. In the case of dynamic scheduling, the system takes care of the work done in the past and that to be done in the future. The fixed scheduling works well in an ideal environment, however, dynamic scheduling takes care of practical constraints under all circumstances.

WORK-ORDER PROCEDURE

For proper scheduling of the maintenance work, the jobs must be controlled through a work-order system, which provides the basic paperwork for planning the workforce, material, and time. A work order is also an authorization to carry out the job. There are two types of work orders normally used in maintenance work—the first type being the standing order for rationalized type of work such as overhauls, repetitive repairs, etc. and the other being the direct work order for non-repetitive types of works such as in the case of unscheduled breakdowns.

CREATING A SET OF PRIORITY FUNCTIONS

Priorities must be established to handle a mixture of backlogged and new pieces of scheduled jobs. The frequency of scheduled maintenance and the criticality of operations are the most meaningful attributes for establishing a priority for a job. Frequency indicates the relative importance of one procedure over the other. If the equipment becomes overdue for a certain number of weeks, it is treated critically important and a critical point is said to have reached. To minimize the critical points, more emphasis should be placed on scheduling the work, when it is due. The jobs should be scheduled according to a priority system. The following priority system should meet the requirements of daily scheduling plans.

Priority I. This priority is used for emergency work. A special work order is used and overtime is also allowed.

Priority II. This priority is used for the jobs scheduled for the next day and some changes in the schedule are possible due to change in priority.

Priority III. Here the jobs can be scheduled depending upon the manpower availability. Once the priorities are fixed, the jobs should be implemented strictly as per schedules for ensuring the proper working operation of the equipment.

It is noticed from experience that maintenance works have a large number of interrelated activities and there will be several jobs in a single department, requiring extensive planning and scheduling by the use of available techniques such as Gantt charts, milestone and network methods.

FORECASTING MAINTENANCE REQUIREMENTS

The generation of schedules for planned maintenance requires a critical study of equipment/systems so that a fair degree of accuracy may be achieved leading to minimization of downtime of equipment, maximization of equipment availability, and therefore, reduction in maintenance costs.

The statistical simulation techniques can be used for forecasting the needs of maintenance. For this purpose, the following breakdown and repair time data of each critical component are carefully analyzed.

(a) A breakdown distribution that indicates the time over which a particular equipment will work without failure.

(b) The repair time distribution to know the time required for a particular breakdown to be completed in order to restore the equipment to its normal working condition.

Knowing these distributions, the breakdown can be forecast through simulation and the repair time can also be known in advance to plan the manpower requirements. These forecasts can be used to choose alternative maintenance policies for each equipment/facility.

PLANNED MAINTENANCE PROCEDURE

In modern industries, the concept of planned maintenance is well accepted since the benefits accruing from it are many. What is needed is the way to take care of planned maintenance and its implementation.

The first step is to establish the number of machines to be maintained. Next, it is to be decided as, to how these equipment/facilities are to be maintained. Then a maintenance schedule is prepared for all critical items, which require preventive maintenance. Initially, this exercise can be started with a few selected components of machines and with the experience gained, it can be extended to other components/parts of machines under consideration.

The main purpose of planned maintenance is to raise the maintenance standards and improve cost effectiveness. This is only possible through a critical analysis of the results of maintenance. This analysis is done with the help of the plant history card, which furnishes the important information regarding the equipment/plant. The information provided by the manufacturer and the technical experience of maintenance personnel in maintaining the plant can be of great importance in analyzing the results of any maintenance work.

EFFECTIVENESS OF PREVENTIVE MAINTENANCE

In order to be effective, the system of preventive maintenance should be able to provide a lot of vital information regarding different aspects of the machine. The efficiency or the success of planned/preventive maintenance programme, therefore, depends on the number of elements needed for this information. Some of these elements are briefly discussed below.

Material Research

Through failure analysis, it can be known whether a particular failure is due to poor quality of the material. In such a case, a thorough investigation of the material is carried out. This analysis can help to improve the quality of the material to the desired level in order to meet the challenges faced during actual working conditions.

Design Analysis

Sometimes the cause of failure can be due to faulty design which can be rectified through redesign or modification of the machine component. In this case, the matter has to be referred back to the manufacturer with the detailed information report regarding the particular equipment, and if possible, the designer can also be called to see the equipment performance at the workplace since some tests may not be possible to be carried out at the laboratory level.

Maintenance Training

It is generally observed in many plants that the maintenance work is attended to on an *ad hoc* basis. In most of the organizations, properly trained persons are not employed for the maintenance work. Considering the basic needs and importance of maintenance, proper training to all the maintenance personnel must be imparted for effective working which will help the maintenance department to be more dynamic and objective oriented.

Operator Training

Sometimes it is seen that untrained operators are assigned the jobs.

The failure frequency of the machines can be considerably reduced through proper on-the-job training of the operators.

Equipment Study

Preventive maintenance programmes, cannot be applied to all types of equipment. The detailed equipment study can, however, identify the critical parts/components, to be included in the preventive maintenance programme. As part of this study, the use of non-destructive testing can also be made of, to identify the critical parts and components prone to defects.

Standardization

Under this scheme, similar types of equipment/components can be effectively maintained in central workshops to provide support to all related departments at one place. It also helps in maintaining better coordination. These shops have testing facilities to examine the equipment/machines critically.

Reports

New developments can be made known to practising engineers through such reports/documents. The reports relating to new maintenance concepts prepared by specially constituted study groups can be useful to the practising maintenance engineers for updating their maintenance knowledge. For example, the concepts of condition-based maintenance and built-in testing are new and appropriate in the present conditions, though they cannot be applied to all types of equipment because of high costs involved.

Protective/Safety Methods

Though the manufacturers normally suggest these methods, it is generally seen that they are not put in practice by the operators/maintenance staff. Strict compliance with these methods can help in minimizing the frequency of failures. For example, the manufacturers recommend the use of distilled water for the radiator of an engine in order to minimize the corrosion of the radiator tubes, but in actual practice it is seen that the operators sometimes do not even use clean water, which increases the failure rate of radiators. Even the protective guards that are provided by the manufacturers are removed by the operators/maintenance staff for their own ease of work, which may lead to some serious accidents.

Work Analysis

For any successful operation it is necessary that the analysis of the working system be made. This helps in identifying the areas where the work is not being done as per requirements, and remedial actions need to be taken in time. It works like a feedback system. In the light of this, an exhaustive analysis of the maintenance programme should be carried out to see its effectiveness and applicability for a particular equipment and environment.

BENEFITS OF PPM

The effective coordination of preventive maintenance work between the operators and the maintenance staff can yield the following benefits.

Reduced Frequency of Failures

The frequency of failures can be reduced by the introduction of a preventive maintenance programme, because the failures can be detected in advance and repair work undertaken before the actual failure. This also leads in reduction of production downtime.

Reduction in Manpower

Under the planned preventive maintenance programme the nature and volume of work is known in advance and therefore it is possible for the management to plan the manpower as per requirements. During inspections (scheduled), plans can be drawn for major works to be undertaken in the future, thus avoiding unnecessary delays involved in the procurement of spare parts and special tools, etc. as required for the work.

Reduction in Downtime

Besides achieving reduction in downtime of equipment, the PPM system also helps in reducing the overtime of the maintenance personnel as the work is executed systematically. Almost all the activities required to be performed are known and arrangements for their execution can be made in advance. As the downtime of the equipment/facility is minimized, the production, therefore, indirectly increases.

Increased Plant Availability

This is closely associated with the reduction in downtime of the equipment. However, in case of a flow process line the failure of one component will put the total system under the down condition. Therefore, preventive maintenance will definitely bring down the failure rate and provide enhanced availability.

Prolonged Equipment Life

It is observed that with the use of a preventive maintenance programme, which includes timely check-up of equipment, lubrication, adjustment, etc. the working life of the equipment increases as the unscheduled breakdowns are minimized. At the same time as the operational running of equipment/facilities becomes better, the consumption of fuel/electricity is also reduced.

Reduction in Maintenance Costs

Though the preventive maintenance programme initially may be costly, but in the course of time the overall maintenance cost of a particular equipment comes down because of proper utilization of the facilities and reductions in unscheduled breakdowns. The frequency of the failures is also minimized, thus, resulting in the reduction in overall maintenance costs. The PPM results in fewer large-scale repairs and fewer replaceable repairs. It helps in better control of spare parts, leading to minimum inventory. Less numbers of standby equipment are needed and thus this system of maintenance assists in reducing the capital investment as well.

Better Quality Control

The PPM is associated with fewer product rejects, less wastage and better quality control because of properly adjusted and maintained equipment. In case of PPM, sufficient time is allowed for maintenance work with the use of specialized test equipment and specialist.

Identification of Critical Areas

The PPM helps in identifying items with high maintenance costs, thus, facilitating investigation and correction of causes such as:

♦ Misapplication/mismatching of parts
♦ Operator abuse
♦ Impact of environmental conditions
♦ Obsolescence

Sometimes the equipment utilization is different from its specifications leading to its underutilization. For example, a heavy duty machine deployed for light work will not be cost effective. In practice, the negligence on the part of the operator also causes some maintenance problems, which can be minimized through proper training. With the development of new technologies, the equipment/ systems go obsolete very fast. Proper attention has to be paid to this issue as well.

To make the maintenance work more interesting from the point of view of the maintenance crew, the concept of maintenance management by objectives can also be tried.

Machine Replacement Time

Having practised the preventive maintenance in any organization it is possible to compare the operating cost with the maintenance costs of the equipment. This information will help the maintenance managers to work out the replacement policy. During scheduled inspection, if the condition of parts/components is noticed beyond repair it would be advisable to replace the same and avoid untimely breakdown.

MAINTENANCE BY OBJECTIVES

It is a management method that can be applied in the area of maintenance as well. The maintenance personnel can select the areas where the improvements in maintenance functions are possible. The incentives for the improvements in maintenance function are decided by the organization and also the areas where the improvements can be possible are well defined. Then the standards for the performance are set and the means of measuring the results are also agreed upon. The objectives may be cost dependent or performance related, depending upon the type of equipment being used. In some areas, services are required round the clock. Under such circumstances, maintenance objectives have to be such that the services are kept in working conditions notwithstanding the cost. Sometimes standby arrangements may have to be made in the interest of continuity of services when the primary system is under maintenance.

The objectives of finance personnel are normally in conflict with those of maintenance personnel. There must be commitment by the maintenance personnel, particularly the maintenance chief, to meet the department objectives in consonance with the financial objectives. For example, the sole objective of the maintenance foreman could be to keep the machine running for the desired production, without bothering about the cost of maintenance. It is necessary to apprise the foreman of management's concern about cost as he may not be willing to curtail any activity under his jurisdiction.

Maintenance by objectives (MBOs) is a real approach to meeting the set objectives. It can be useful in evaluating the performance of a foreman, motivating the personnel to make commitments to achieve satisfactory results with their own objectives. To obtain motivational value from an MBO programme, the maintenance chief must be able to spell out his department objectives and then clearly communicate them to his foreman and other related persons. For example, the manager may set an objective of reducing the labour cost by five per cent. This objective may be based on the company objectives to conduct maintenance at a fixed rate per unit produced. Since this reduction is in order, the manager must endeavour to improve the existing system to reduce the labour cost by five per cent. Once the manager has made his objective clear, he must then guide the foreman to achieve this goal by reducing the overhead expenditure or by improving the performance of the maintenance personnel.

The idea of the foreman on reduction of labour cost must be given due weightage because of his direct involvement in the problem. His approach may be different from that of the manager/engineer. The foreman must state his objective in terms of the results expected by him. When the objectives are achieved, the foreman must be commended, and if he fails, the engineer must do counselling, who must remember that he is the one who is responsible for everything that his foreman does.

Management by objectives (MBOs) can also be an effective tool for sharpening the operating skills of maintenance personnel. This is only possible when there is close cooperation between the maintenance engineer and his staff. The improvements in the performance of maintenance personnel can be assessed with the help of MBOs, taking care of the following points:

- The attitude of the staff to improve themselves and how the maintenance engineer can help them to achieve their goals.
- What level of guidance is required? Is the staff being asked to do more than what really they can do?

Did the staff achieve their objectives as per schedule?

The same individuals who have to achieve objectives must prepare the MBO programme, because objectives without a plan are useless. Therefore the active participation of the maintenance staff is essential during the planning stage. When the programme has been put under operation, the foreman and the engineer should review the progress of the work periodically. The foreman must prepare the report indicating the progress or shortcomings of the programme.

During review, if the foreman feels that the goals set by him are not realistic, they may be changed in due course of time to achieve the final objectives of the maintenance programme. If the progress is not satisfactory as observed during review, the foreman has to re-orient his efforts properly. He must keep in mind that planning and scheduling are an integral part of the management procedures. The plan of action for fulfilling the objectives must be communicated to all concerned persons and their cooperation must be sought.

For successful completion of maintenance by objectives, the following steps must be followed throughout the life of the programme.

1. List out the common goals of the maintenance department and specify the responsibility with accountability for each of the goals.
2. Specify the role of each and every maintenance person so that they can be held responsible for any deficiency in completion of the MBO programme.
3. To meet the goals/objectives, commitment to targeted dates is important for all. For example, the preparation of proper schedules for every job by a targeted date.
4. If the targeted dates for the completion of jobs have slipped, a programme for review of the work must be drawn.
5. When the objectives are met, it must be made known to all the concerned persons and the key persons must be rewarded for their achievement.

The main problem faced by any engineer today is the control of labour. Setting the goals, and providing certain motivations for improving workmanship can minimize this problem. In most of the cases, the supervisory level plays an important role in improving the existing levels of performance of the workers. A better co-ordination with all the concerned persons yields fruitful results and therefore should not be neglected at any stage.

DEVELOPMENT OF CHECKLISTS

Generally, the manufacturer, along with the machine, supplies the checklists as well. They also modify and update the checklists during the guarantee period when they take up the different maintenance jobs. Whether a particular equipment or system will be brought under preventive maintenance and whether some items of the equipment or system would be brought under checklist inspection, depend on a number of factors such as those enumerated below:

- ◆ Criticality of the item
- ◆ Availability of standby equipment/facility, i.e. redundancy
- ◆ Cost analysis
- ◆ Expected service life
- ◆ Technology advancement
- ◆ Rapidity of slipping into obsolescence.

While developing the checklists, the following steps need to be considered:

1. The total number of subsystems that the machine or system is comprised of

2. Analysis of the functioning of these subsystems and identification of their critical components

3. Identification of the indicators of functional behaviour of the critical components. These may include:

   ◆ Abnormal sound
   ◆ Vibrations
   ◆ Leakage of fluid
   ◆ Wear
   ◆ Cleanliness
   ◆ Play
   ◆ Cracks
   ◆ Change in colour
   ◆ Corrosion
   ◆ Temperature of bearings
   ◆ Electrical failure
   ◆ Performance of electrical connections
   ◆ Power/fuel consumption

4. Determination of the normal, alarming and dangerous levels of behaviour of components/equipment, etc.

5. From previous experience, determination of the safe period of withstanding the abnormal behaviour

6. Determination of the periodicity of inspection of the various indicators. This depends on:

   ◆ Age and condition of the asset/equipment
   ◆ Severity of service
   ◆ Safety requirements
   ◆ Hours of operation
   ◆ Susceptibility to wear
   ◆ Susceptibility to damage
   ◆ Susceptibility to deviations from normal adjustments

7. Development of a proper format for conducting the checks and formulating the reports.

IMPROVING SKILLS OF MAINTENANCE PERSONNEL

It is observed from the above discussions that by the use of a planned preventive maintenance programme, the availability of the equipment/facilities can be improved. This also helps indirectly in improving the reliability of equipment/system at the same time, better utilization of the available manpower can also be made. The right type of person if assigned the right type of job will improve the skills of maintenance worker. An important parameter of the maintenance programme is its evaluation, which will be discussed, in the next chapter.

QUESTIONS

1. Briefly explain the objectives of planned preventive maintenance.

2. What is the scope of preventive maintenance in an organization?

3. Explain the elements of a planned preventive maintenance programme.

4. Indicate the areas where the planned preventive maintenance methodology can be applied effectively.

5. How is work planning and scheduling carried out in a maintenance organization?

6. How can the effectiveness of preventive maintenance help the maintenance department?

7. Discuss the benefits that accrue from planned preventive maintenance.

8. How does maintenance by objectives help any maintenance system? Briefly explain the factors involved.

9. How can the skills of the maintenance personnel be improved? Discuss.

10. How does a planned preventive maintenance system help in identifying the critical areas of maintenance functions? Explain.

11. How will the development of checklists help in improving the maintenance functions? Justify your answer.

12. Suppose you are the chief of a maintenance department of a cement plant. Discuss your plan for implementing maintenance by objectives.

Chapter

3

Maintenance Evaluation

INTRODUCTION

The role of the maintenance department may become pivotal in assuring the economic competence of any project where heavy duty and costly equipment/systems are deployed. The objectives of different types of maintenance and their roles are described in the previous chapters. In this chapter, the importance of maintenance evaluation and the methodologies thereof are highlighted.

F.W. Taylor and F.B. Gilberth are considered to be the pioneers in the field of work measurement, using time and motion studies. Their studies are normally applied to production work. Now with the advancement in technology it has become necessary to measure or evaluate the maintenance work as well. Since maintenance cost is increasing day-by-day due to sophistication of equipment and degree of accuracy required, it is of prime importance to employ effective measures for maintenance jobs to improve equipment availability and achieve reduction in maintenance costs.

The success of an organization depends upon the effective utilization of its manpower and resources. The objective should be to utilize the potential of each employee for the benefit of the individual as well as that of the organization. The most important consideration in this regard is to identify the parameters that will measure the maintenance performance. These parameters would be used as maintenance evaluation index (MEI) and be of much value as tools of any decision-support system for project management. A basic relationship between the maintenance team performance and production cost helps to act as a decision-support tool in this regard and can be used to make better assessment of maintenance functions.

To decide the structure of wages and salaries, job evaluation has proved to be an effective aid in providing an equitable and supportable basis. It is also necessary to keep the general level of compensation in line with competition, nature of job, and the geographical area in which a particular organization is set up. First of all, the jobs must be classified based on job content and characteristics for determining the rate of payments. A committee consisting of trained and qualified persons may compare the jobs for the purpose of evaluation. Every job has certain variable factors such as skill, responsibility, application, and working conditions. Unskilled persons may undertake some of the jobs while there are many other jobs, which require skilled and experienced persons. After a thorough comparison of jobs, the key jobs may be selected that need wide requirements of skill, responsibility, and other important factors. These key jobs can be used as a guide for all other jobs.

The next step is to ensure the description of the important jobs to give them the degree in terms of the factors such as education, experience, and leadership involved in each particular job. This will help to assign an index number to each category of operations for job evaluation. Depending upon the nature of operations in order of their relative difficulty the jobs may be ranked. The index values, thus obtained for each operation, will also help to fix the rate of wages. The nature of maintenance work will also play an important role for the fixation of salaries to the maintenance personnel.

Maintenance evaluation can be defined as an assessment of successes and failures in maintenance techniques. The resulting plan is used to determine the financial and staffing requirements to complete the tasks. This data can be used to determine equipment, tool and vehicular requirements as well as the potential for volunteer and service club involvement.

BACKGROUND OF MAINTENANCE FUNCTION

Before actually proceeding to the maintenance evaluation process it is essential to know the background conditions in which the system has to operate. There are various challenges to be faced by the maintenance personnel. The maintenance function is still one of the most neglected areas in many of the organizations even today. In our country, there are still organizations that do not have proper manpower and equipment/tools to perform the maintenance function. Hence the planned/preventive maintenance is often overlooked in many cases and due attention is not given to the maintenance function which can improve the availability of production/service equipment. The maintenance evaluation process should therefore be applied where separate organizations are set up and the maintenance work requirements are high. It is also essential where the equipment/ services are complex in nature and their use calls for safety and accuracy. The use of evaluation under these conditions will help proper deployment of maintenance personnel.

NEED OF EVALUATION

The need of maintenance evaluation can be estimated by comparing the results obtained with and without the work measurement programmes. By work sampling, the causes of poor performance and the possibility of improvements can both be identified. Due to the involvement of more maintenance personnel in modern organizations, the need of maintenance evaluation has further increased. The main objectives of maintenance evaluation are:

◆ Avoiding overstaffing and understaffing
◆ Setting the standards for different types of works in advance
◆ Evaluation of maintenance function through indices
◆ Type of maintenance needed for a particular equipment/ system
◆ Spare parts requirements
◆ Skill of maintenance personnel required

These evaluations will also indicate the level of maintenance existing in various organizations. They will give a fair idea about the overall performance of the organization. The performance analysis at different levels starting from the foreman to the maintenance engineer can also improve the work culture. This will give the organizations/departments a clear picture, as to where they stand in the fast-changing technical environment in order to be able to carry out effective improvement of the maintenance function. However, maintenance functions are affected by numerous variables. Therefore, it is experimentally difficult and complex, if not impossible, to measure the effectiveness of maintenance in absolute terms. Poor housekeeping, shabby appearance, and lower machine life are all indicative of bad maintenance.

Effectiveness of maintenance is generally reflected in good housekeeping and better equipment conditions. Many industries consider machine availability as a ratio of the hours worked to the sum of the worked and repair hours. In measuring this machine availability, no allowance for waiting for spares is given, since inventory planning and management is also a job of the maintenance department. Reduction of unscheduled downtime and alternative use of maintenance downtime are also reflective of better maintenance performance. Hence these parameters should also be analyzed while evaluating the maintenance performance.

MAINTENANCE FUNCTION REQUIREMENTS

Before actually starting the evaluation of maintenance work, the challenges likely to be faced by the maintenance organization should be anticipated in advance. This work requires the following information:

1. Mechanical requirements of the equipment/system to enhance its availability.
2. Requirements of civil engineering works for new facilities and infrastructure to facilitate maintenance.
3. Works imposed by requirements of environmental protection and pollution control. This includes maintenance of oil traps, dust, and exhaust systems, and so forth.
4. Electrical requirements for better utilization of available energy.
5. Failure analysis of the equipment and assessment of reliability for providing feedback information in respect of the concerned equipment and system design.
6. Customer satisfaction while using facility in terms of comfort etc.

BENEFITS OF MAINTENANCE EVALUATION

The benefits that can be achieved through proper work measurements are as follows:

Performance Improvement

The main benefit is generally improved maintenance performance and a decrease in labour cost. Without real assessment of the maintenance work, the performance will be poor. With the application of the evaluation methods, the goals/targets for each individual can be well defined which can boost the morale of the working people and improve their performance.

Reduction in Delays

If the work measurement control is applied to maintenance function, the delay in performance of jobs can be reduced. It can also give reasons for delays; sometimes a good percentage of time is lost in the procurement of maintenance materials. Such delays can also be minimized. It is observed in a number of cases that the maintenance work suffers due to non-availability of spare parts. A proper evaluation will help identify the reasons for delays in the maintenance function and then the steps required for the improvement can be taken.

Reduction in Equipment Downtime

The downtime of the equipment can be reduced through maintenance evaluation. When a machine is out of service and put under repair, the use of standard evaluation will bring out better labour performance. It also becomes possible to calculate the actual downtime of equipment and thus standby systems can be used, if required. Moreover, the use of standard evaluation helps in planning the scheduled maintenance and tools required for the work.

Improvements in Preventive Maintenance

Work evaluation requires detailed analysis of the practices used, their correctness, and fault corrections needed to be applied to different maintenance systems. Differences, if any, from standard requirements can be clearly identified and corrective measure can then be taken to reduce the maintenance costs. Further, by studying the current practices, the preventive maintenance system can also be improved. For example, an alternative solution can be thought of beforehand if an undue delay in procurement of spares is known to be taking place due to some problems at the supplier's end. The requirement of spare parts can be identified during the evaluation process and critical items can also be known in advance.

TYPES OF EVALUATION

The evaluation of the maintenance function can be carried out by comparison with the similar type of work. If information about the same type of work is not available from elsewhere, then the evaluation of the maintenance function can be performed over a period of time. For effective evaluation of the maintenance function, it is desirable to carry it out at different levels within the maintenance department. A three-tier evaluation system is possible, the first being at the foreman level. Here it may include effective utilization of maintenance crew and the total time taken for the repairs/overhaul, machine-wise or category-wise. This evaluation will indicate information about the maintenance work planned, actual work completed, backlog, etc. Proper indices for backlog repair/maintenance help to produce effective utilization of the available manpower. This in turn can initiate measures for improvement.

The second and most important level for maintenance evaluation is at the level of the deputy manager who is responsible for guiding the foreman in the matters of control of manpower and the costs involved in maintenance work. The important parameters for performance evaluation may include labour productivity, availability of production machines/facilities, time, cost, materials, and spare parts needed to do the job. Each of the above parameters needs to be critically analyzed to find out the reasons for getting the desired results from the maintenance function.

The last level of evaluation is at the level of the maintenance manager. The indices developed help the maintenance manager in evaluating the performance of the maintenance department as a whole and enable him to draw up plans/programmes for improvements to the maintenance function. At this level the analysis of the evaluation done at the foreman level and at the deputy manager level is carried out to develop maintenance indices. Different types of maintenance evaluation techniques are now described below.

Evaluation through Reports

The evaluation of the maintenance function can be based on reports which are prepared at fixed time intervals, i.e. weekly, monthly and yearly, giving details of the equipment serviced/maintained during the said period. The accuracy of the reports will reflect the real evaluation of the process. The reports should give the complete picture of the maintenance function and should not be treated as a mere formality, so that the corrective measures can be taken. Such practices if followed for the maintenance function will improve the total performance of the organization. Such reports also help other organizations, which use the same type of equipment. If a particular type of failure is reported by all the organizations, it may call for a modification or a design change in the equipment. The impact of environment on equipment/system can also be known through the reports made available.

Subjective Evaluation

No evaluation can be done in isolation, since this has to meet organizational objectives and should be consistent in nature. The norms set for the maintenance function evaluation should be valid. Some of the characteristics of subjective evaluation can be expressed through numerical values, but the others can be qualitative estimates such as good, fair, adequate, and so forth. Here, the parameters of evaluation will depend upon the available expertise on the subject, qualifications of the working people and their training, since it is expected that the better-trained people would perform the job in a rationalized manner. Adequate information about the work can also help in the preparation of related documents to improve the availability of the equipment/service.

Objective Evaluation

A number of indices are frequently used for objective evaluation. These include:

$$R1 = \frac{\text{Mean time between shutdowns (MTBSD)}}{\text{Mean time between failures (MTBF)}}$$

$$R2 = \frac{\text{Mean time between failures (MTBF)}}{\text{Mean time between repairs (MTBR)}}$$

$$R3 = \frac{\text{Equipment availability}}{\text{Maintenance cost}}$$

$$R4 = \frac{\text{Time used to repair}}{\text{Maintenance cost}}$$

$$R5 = \frac{\text{Number of breakdowns in a given period}}{\text{Total time of equipment availability}}$$

$$R6 = \frac{\text{Mileage covered in given time}}{\text{Total maintenance cost}}$$

$$R7 = \frac{\text{Maintenance expenses per tonne of material produced}}{\text{Total material produced in given time}}$$

$$R8 = \frac{\text{Maintenance costs}}{\text{Production costs}}$$

Though these ratios may vary from one organization to the other, they can be used for comparison. The organizations can develop their own ratios and norms for evaluation purposes, keeping in mind the major objectives of the evaluation such as equipment availability, reliability, and so forth. The ratios given above can also be helpful in assessing the maintenance costs, which are the key factors for the containment of the system being practised.

The evaluation of the maintenance function can also be carried out by asking questions related to maintenance, such as what type of system is being followed in a particular organization. This will give an idea about the importance attached to the maintenance function at various levels, however it will be far away from the real life situation.

The ratios defined above can be computed on monthly or yearly basis, which can also be treated as being indicative of the performance of the maintenance personnel. The basic objective of maintenance evaluation would be to find out the problems being faced by the management.

Job estimations are used in some cases for giving incentives to maintenance personnel. When this system is used, the time for every job is estimated through a time-and-motion study, and a standard time is set for every job. This helps the management to evaluate the performance of the working personnel. Those who complete their jobs within the standard time are paid extra money for the extra job completed. On the other hand penalties are imposed on the workers who do not complete their work in time.

The planner bases the formal planning and scheduling of systems on time estimates. Since time estimates are based on existing levels of work performance, true labour utilization cannot be predicted and performance cannot be evaluated based on the attainable levels of performance. These estimates are normally inconsistent in nature, but being less costly, this approach is used almost in every organization.

To reduce the downtime of equipment during repair, it is essential to follow some method to find out the repair time required for a particular job, which will help in manpower planning. These standard times can be more useful in the case of repetitive type of work such as overhauling of components. The task of job estimation is difficult for the non-repetitive types of jobs or for the jobs of specialized nature. However, the experience of working personnel can prove to be very useful in estimating the approximate time of such jobs as and when required.

Work sampling is widely used for measuring the effectiveness of maintenance personnel. Here the observer records the idle time as well as the percentage performance of the workers while they are working. This information may be used to assist in improving the maintenance work methods as well as in achieving reductions in idle times and delays in the performance of the maintenance work.

This evaluation also offers an excellent means for making preliminary surveys on labour utilization. The accuracy depends upon the observations taken, because sometimes the workers may be keeping themselves busy without actually doing any useful work. It cannot be used as a continuous measuring tool because once the workers know about it, they will try to confuse the sampling techniques. The following information obtained through work sampling can be utilized to control:

1. Total utilization of the group.
2. Percentage of time spent on various activities to identify the areas of improvement.
3. Utilization of available facilities.
4. Requirements of special testing facilities for maintenance functions.

If the individual work observations are recorded and rated, it is then possible to obtain an overall effectiveness index, which can also be used by other organizations for comparison purposes.

Work Auditing

This method is slightly different from work sampling as it provides the true cause of delays that result into high job costs. In this case, the step-by-step progress of the randomly selected jobs is recorded to highlight the delays caused. The delay may be because of poor planning of the repair work being undertaken or the non-availability of spare parts and the special testing facilities required.

An analysis of the total time for which the equipment is not available for production work and the total time taken by the maintenance staff to attend to the fault, would enable the maintenance foreman to understand the weak points with a view to getting better results. The areas where improvements are possible can be identified and corrective actions initiated accordingly.

Depending upon the various working conditions such as equipment location with respect to the maintenance section, average waiting time, number of jobs/equipment in waiting and so forth, suitable realistic indices can be developed to measure the performance of the maintenance personnel.

STATISTICAL ANALYSIS OF PERFORMANCE

The information that is available from the past can be analyzed for the purpose of fixing the indices for maintenance evaluation. Data may be considered for a period of six months or more. The jobs are classified under various categories such as:

- ◆ Standard orders and routine work
- ◆ Repetitive jobs
- ◆ Estimated jobs
- ◆ Miscellaneous jobs
- ◆ Emergency jobs

For the purpose of analysis, standards are developed, which are used in maintenance scheduling and planning. It is a simple method of work measurement but not accurate enough to give consistent work performance figures. However, it is good for routine or repetitive types of jobs.

The analysis of the total time spent on the repair of equipment can be done to analyze delays. Such delays may be due to:

- ◆ Lack of proper instructions to the personnel
- ◆ Waiting for spare parts/materials
- ◆ Shortage of tools/special equipment
- ◆ Work held up for want of power supply
- ◆ Non-availability of qualified maintenance personnel

Such analysis helps in knowing the time lost in various undesired activities so that the same can be minimized with appropriate actions.

Some plans are prepared on the basis of statistical analysis. The factors used include time allowance for routine and repetitive jobs as well as downtime, machine utilization, percentage capacity, etc. These help in evolving standard times or allowances based on past records, which may however be modified as per the requirements from time to time.

Job Slotting

In this type of analysis, jobs are selected to develop standards in order to determine the job completion time for similar types of jobs in the future. This method yields accurate results since it is based on past data, though the initial cost may be high. The skill of the worker is also of prime importance for collecting the information. Here bigger jobs are divided into smaller components to critically examine them and make effective evaluation.

Standard Data

This system provides a more accurate and thorough method of work measurement in maintenance applications. It may be costly and time consuming but gives effective maintenance control. The basic problem of this system is to develop standard data, which can only be done by the use of actual time-and-motion study observations. The activities are classified as productive work, indirect work, idle time and travel time for each job, and time estimates are made for each activity. The information thus obtained will help to identify the areas where improvements will be possible.

SELECTION OF WORK MEASUREMENT METHODS

There are several factors that affect selection of the work measurement methods such as purpose, control, and the incentive programme. The selection of such methods depends upon the number of persons employed and the nature of the incentive plan. The work methods cannot be used effectively until the organizational structure cooperates to implement them. Before introducing any method of work measurement its possible merits and demerits are therefore analyzed, otherwise it may prove to be a failure. However, these work maintenance methods can be an aid to good maintenance management.

COST OF MAINTENANCE EVALUATION

The implications of maintenance evaluation to the organization is an undesired activity and include the following costs which also need scrutiny for the benefit of maintenance department:

1. The one-time cost of personnel employed for the maintenance evaluation work. This work can be better done by an external agency.
2. Idle time cost of the maintenance workers for answering the queries of the evaluation team.
3. The increased recurring maintenance costs due to revision of the wage structure on account of the introduction of the new job evaluations.
4. The cost of maintaining the job evaluation team as part of the maintenance department for evaluations from time to time.

Therefore, it is desired that before adopting any of the above techniques for the job evaluation, all the related parameters must be examined in detail. Otherwise the exercise may prove to be a failure and the organization may face the adverse reaction of the trade unions.

Example: The following chart shows the failure history of a machine. Determine its MTBF.

Failure event	Cause-wise duration of failure (min)					
	Time	1	2	3	4	5
1	8:30	5				
2	9:40			15		
3	11:15		10			
4	13:30					30
5	15:20			8		
6	22:00				12	

Solution: There are five intervals of failure. The durations are given as:

$$8:30-9:40 = 70 \text{ min}$$
$$9:40-11:15 = 95 \text{ min}$$
$$11:15-13:30 = 135 \text{ min}$$
$$13:30-15:20 = 110 \text{ min}$$
$$15:20-22:00 = 400 \text{ min}$$

Thus the total time between failures is 810 min. Therefore,

$$\text{MTBF} = \frac{810}{5} = 162 \text{ min}$$

QUESTIONS

1. Why is maintenance evaluation important for successful working of the maintenance function? Discuss briefly.

2. Describe the benefit, which can be derived from maintenance evaluation.

3. Discuss the methods employed for maintenance evaluation.

4. Discuss how statistical analysis can help in improving the maintenance function.

5. Briefly describe how the selection of a maintenance evaluation method can be made.

Chapter

4

Condition Monitoring

INTRODUCTION

Condition monitoring is a technique, based on the advanced technologies. It is used to determine the condition of equipment and predicting the failures with maximum success. It includes but is not limited to technologies such as:

- Vibration measurement and analysis
- Infrared thermography
- Oil analysis and tribology
- Ultrasonic
- Motor current analysis

With technological growth, the design and development of high-speed machinery is possible, which calls for their effective working without any intermediate interruptions. The high-speed machine is subjected to vibrations and this calls for vibration analysis. Although, strain, force, pressure or temperature, occasionally need to be measured for rotor dynamic evaluation, the two most commonly measured variables are vibration and rotor speed. The availability of such equipment should be high for their better utilization, which is only possible through advanced maintenance techniques. One out of them is condition monitoring, where components/systems are checked at regular intervals to avoid impending failures. This leads to undertaking corrective measures when they are needed rather than at the scheduled or routine intervals. Thus, it can minimize the downtime of equipment for unnecessary inspections. Most of the processing and power industries utilize some form of condition-based maintenance for higher reliability of their equipment/system.

Figure 4.1 shows the applications of condition monitoring in an automated manufacturing/processing organization. With the help of condition monitoring, it is possible to prepare a maintenance database, which can be utilized for condition-based maintenance planning

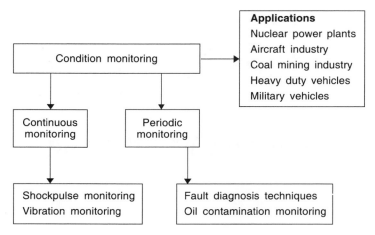

Figure 4.1   Applications of condition monitoring.

and control. It also contributes to maintenance planning, maintenance cost reduction, health and safety programmes with savings in energy by providing early warnings of waste and inefficiency arising from faulty operation. The work of maintenance planning is facilitated by advance appraisal of defects through condition monitoring. Such pre-planning also helps to extend support to several aspects of maintenance work such as, spare parts, technical information, and man-power requirements. It also helps in completing the maintenance work in scheduled time, thereby effecting reductions in overtime payments.

Here the conditions of equipment/system are compared with the pre-set acceptable standards which are established during installation and commissioning. This is considered as one of the reliable and cost effective maintenance techniques for capital-intensive equipment/machinery. Under certain circumstances it is the compulsion for the users for reasons of human safety. Thus, it is important and necessary to develop an appropriate monitoring system in the overall maintenance schemes for all machinery in order to optimize the maintenance costs.

Monitoring techniques not only enable equipment to operate at optimum performance with improved reliability but also provide information on trends in machine deterioration to serve the purpose of scheduling constructive maintenance without affecting production.

BASIC CONCEPT

The industries all over the world today demand high-technology maintenance management in view of the high **power and speed** of machines utilized in the complex processes/terrain conditions. Minimum downtime of machines and reduction in penalty costs are the paramount needs of every enterprise. These considerations are particularly true in the areas of advanced technology such as aircraft industries, oil well drilling, coal extraction, power plants, and chemical industries, etc. Condition monitoring has to achieve the maintenance objectives in such highly capital-intensive industries.

Condition monitoring is the process that assesses the health of an equipment/system at regular intervals or continuously and exposes incipient faults, if any. Basically, it consists of extraction of information about particular parameters from machines and analysis of data to predict the health of machines without affecting their operation. Condition monitoring leads to undertaking corrective measures only when they are needed rather than at the scheduled or routine intervals. Thus it can eliminate downtime for unnecessary inspections.

This kind of predictive maintenance has proved to be very successful in the capital-intensive industries. Most processing and power industries utilize some form of condition monitoring for higher reliability of their equipment. Normally, regular inspections are carried out on the parts/components at pre-specified intervals. The wear and tear or failure of the part are detected during such inspections. Condition monitoring involves application of fault diagnosis techniques. A number of such techniques are available to identify the causes of impending trouble in a machinery.

Using diagnostic instruments on a routine or continuous basis is the way to carry out routine condition monitoring. The performance and the condition of machinery are monitored and compared with acceptance standards. The acceptance standards are established during the commissioning tests. In such tests, the parameters are measured against known loads or predetermined conditions. The statistical analysis of the detailed breakdown information has revealed that one-third of the total available time is lost due to breakdowns.

Condition monitoring is considered as the most reliable, cost effective and efficient technique for maintaining the majority of critical equipment such as engines, turbines, compressors, etc. used in most of the industries today. Approximately 20–30 per cent of possible production time goes towards maintenance of equipment and this can be minimized through effective maintenance. Towards this aim, condition monitoring plays a very important role. Thus it is necessary to develop an appropriate monitoring system in the overall maintenance schemes for all machinery.

Of the two types of condition monitoring, namely periodic and continuous as shown in Figure 4.1, the former is normally opted for non-critical machines. For the cost-intensive critical systems or machines, continuous condition monitoring has proved to be highly cost effective.

LEVELS OF CONDITION MONITORING

Condition monitoring can be carried out at different levels. As shown in Figure 4.2, there are four distinctive levels.

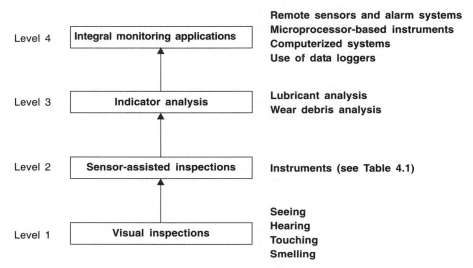

Figure 4.2   Levels of condition monitoring.

The level 1 (visual inspections) form part of the normal preventive maintenance and is generally included in the daily and weekly maintenance activities. The persons involved are expected to sense the condition of equipment by seeing, hearing, touching, and smelling. Magnifiers, viewing devices, temperature sensors and other instruments can further assist in this process.

At level 2 the persons are assisted by the available portable instruments to make various measurements. For example, Table 4.1 shows the instruments that can be used for this purpose.

Table 4.1   Instruments used for condition monitoring

Measurement	*Instruments*
Speed and distance	Tachometers, Odometer
Electrical quantities	Test meters
Fits and clearances	Proximity meters
Temperature	Thermography, Pyrometers
Wear	Thickness gauges
Corrosion	Corrosion meter
Movement	Vibration analyzer, Frequency analyzer

The condition of the lubricant, or the presence of wear debris in the gear box oil or engine oil can be used as indicators of the condition of the system. However, this condition cannot be assessed by mere human senses. Advanced techniques for analyzing the lubricant for its condition or contaminants are now available. The metal particles present in the engine oil can indicate the condition of the engine cylinder or piston. Therefore, the contaminants in the lubricant provide clear indications of the machine condition. In the level 3 condition monitoring, the analyses of these contaminants are carried out. For such analysis, a sophisticated test laboratory is necessary.

The level 4 monitoring can be carried out with the help of mini-computers or microprocessors. Here, process transducers, accelerometers, counters and other sensors are employed at different points of the equipment for acquisition of data to indicate the failure of the system. The use of data loggers and computer networking and data transfer have all revolutionized the level 4 condition monitoring systems. However, level 4 condition monitoring involves high costs and skills and thus it is suitable for large-scale industries only.

CONDITION-MONITORING TECHNIQUES

In the case of mechanical equipment/systems, the most vital and significant components/parts that need regular inspection are support bearings. Failure of these parts leads to excessive vibrations, which in turn propagate other related problems and cause subsequent premature breakdown of machinery. Monitoring of vibrations, therefore, is essential to study the actual condition of bearings. An analysis of machinery vibration and noise should also be made at the start of a preventive maintenance programme to determine whether or not the machine is in good operating condition. If faults are present in a machine, the initial analysis will indicate these for corrective action to bring back the system to its normal working condition. The analysis data taken with the machine in good condition is called the 'base line data' and provides the basis for comparison with future periodic checks and analysis.

The success of condition-based maintenance depends on the efficiency of identifying the deteriorating trend in the machine components. For this purpose it is essential to recognize the sources or causes of failure. Failures of different types induce different types of effects on the system. These effects may be classified as dynamic effect, particle contaminant effect, chemical effect, physical effect, temperature effect, and corrosion effect. The condition-monitoring system aims at recognizing these effects and assessing their degree with the help of some specially designed devices. These devices or the instruments provide a quantitative measurement of the effects and thus help to assess the condition of the machine/system. Some of these effects and their assessment are briefly discussed in the following sections.

Dynamic Effects

Dynamic effects are those, which can be identified through vibration analysis, pulse monitoring, and acoustic emission analysis. The analysis of vibration signals, produced during the operation of the machine, provides important information about the condition of the machinery. Vibration analysis can identify the machine problems like imbalance, misalignment, mechanical looseness, bent shaft, bad bearings, gear damage, electrical trouble, or faulty aerodynamic behaviour. Undue vibration levels will eventually produce serious faults and if detected in time, a suitable repair or adjustment scheme can be implemented. The analysis of vibrations enables the maintenance engineer to conduct routine checks on rotating parts to keep the vibrations within limits. The individual component vibration frequencies or the total system vibration level can be measured with a stroboscope and the images of the vibrating parts can be had on an analyzer. The meters indicate acceleration, displacement, or velocity of the parts.

Vibration signature

Some of the typical active systems have their own signatures during operation. These include internal combustion engines, pumps, fluid valves and bearings, etc. Vibration signatures or the patterns of vibration during operation can be divided into three groups depending on the source of vibration. These are as follows:

1. Cyclic machinery (e.g. engines and transmission systems)
2. Flow noise generators (e.g. pumps, boilers)
3. Single transient generators (e.g. switches, punches)

It is noticed that cyclic machinery produces a sound or vibration pattern which repeats at a certain time interval, and this can be analyzed for the purpose of condition monitoring. The changes in the signal will indicate the problems associated with the system that can be analyzed for a particular defect incurred in the parts of the system. The frequency of vibration is the most important characteristic as it can indicate the cause of vibration. Vibration analyzers are available, which can measure and indicate the amplitude, frequency and phase of vibration. Such analyzers are normally very sensitive and can capture amplitudes of less than a micron. Normally, there are recorders that give printout of the vibration characteristics from their own in-built printer or the data stored can be downloaded to a PC for further analysis. It is desirable to use the selective methods to reduce the monitoring cost and also to know the kind of the failures likely to occur. The single-point signals do not give good indications regarding the condition of machine. Therefore, measurements should be made at different points and in different directions and be properly studied to understand the complex waveforms so obtained. The selection of points and direction for measuring the signals is a very important and perhaps the most critical factor. If this selection is not done properly, no amount of analysis, however thorough, will reveal the correct condition of the machine.

The best locations for vibration measurements are the bearings, where the forces and loads are applied. If it is not possible to measure the vibrations on the bearings, then the same can be measured at locations as close as possible to the bearings. For a complete vibration signature of a machine, three-dimensional measurements should be made at each location, as the different malfunctions cause vibrations at different discrete frequencies. The combination of these discrete frequencies create a complex waveform at the point where the measurements are made. These measured signals have to be analyzed and presented as a plot of amplitude and frequency, which is referred to as the vibration signature of a machine.

In the spectrum analysis, recordings are made at regular time intervals and compared with the previous recordings or original recordings, which are used as a standard. Attention should be given to those frequencies that do not match with the original frequencies with a view to finding the faults in the system.

To introduce a vibration monitoring based maintenance programme, the following arrangements should be made:

♦ List the critical machines to be included in the programme.
♦ Establish the acceptable level of machine vibration.
♦ Determine the normal vibration characteristics for each machine.
♦ Select, identify the periodic checkpoints for all the mechines under the programme.
♦ Estimate the periodicity of vibration checks for the selected checkpoints.
♦ Design a simple data recording and storing system.
♦ Develop an adequate data analysis procedure.
♦ Train the personnel to carry out the vibration monitoring and analysis.

Vibration monitoring equipment

Transducers, vibration pick-ups or strain gauges detect the vibration signals, which pass through a machine component to the machine structure. In its simplest form, the vibration monitor may present a safe/unsafe indication of vibration severity, whereas an advanced vibration monitor may provide a spectrum or signature by means of which variations in the discrete frequency of the defective component can be used to identify the component which is likely to fail. In some cases, the data can be processed in detail to find out the type of defect within the component itself.

Vibration analyzer

The vibration analyzer is used to pick up the signals from the working system in order to study the condition of a component/machine. With the help of a vibration analyzer, these complex waveforms can be analyzed by means of narrow band filters. The frequency range obtained enables the maintenance engineers to prepare routine check-ups on rotating machinery/parts and to know that the vibration frequencies are within the specified limits. In a complex machine, the individual component vibration frequencies as well as the overall vibration level can be measured. The meters, provided in the vibration analyzer enable to measure acceleration, displacement, or velocity.

Octave analyzer

The octave analyzer is used to measure the amplitude and frequency of components in a complex sound and vibration spectra. The analyzer consists of a high-impedance amplifier, a continuously tunable filter with narrow bandwidth. The centre frequency of the filter is continuously adjusted.

Real-time analyzer

An online real-time analyzer provides the analysis of the frequency spectra. Such analysis permits real-time processing without any loss of data. The use of programmable microcomputers have made it possible to develop these instruments incorporating a broad range of analysis techniques for the time series data. The other instruments used for the purpose of vibration analysis include a percentage bandwidth analyzer, a narrow bandwidth analyzer, a turbine vibration analyzer, and vibration limit detectors. These instruments indicate changes in vibration frequency and enable detection of shock pulses and plastic deformation.

Particle Contamination Effects

These effects are associated with failures due to wear or fatigue in which discrete particles of different shapes and sizes are emitted into the working medium, e.g. hydraulic fluid or lubricating oil. Through quantitative and qualitative monitoring of contaminating particles and deterioration of the carrier liquid itself, it is possible to assess the extent and source of contamination within the system. The wear particles are generated in engine transmission, gear box, hydraulic systems, turbine and compressor lubricants. Detection of wear particles can assess the rate of wear in order to take mitigating measures. The techniques used for this purpose include ferrography magnetic chip collector, X-ray fluorescence analysis (XFA), graded filtration, and blast testing. Wear fatigue and corrosion can be detected by the use of these techniques which in turn help maintenance engineers to propose schedules for checking the lubricants and hydraulic oils from time to time to avoid damage to the components.

The main contaminants that could be present in the lubricants or hydraulic oil are aluminium, iron, antimony, lead, silicon, chromate, silver, copper, tin, etc. Aluminium particles can be found where aluminium pistons are used and such particles indicate wear of piston. Iron contamination comes from the use of piston liners and piston rings. It becomes more prominent in the presence of combustion gases due to burning of the oil film. The copper and lead particles are mostly found in the places where bearings are made of these materials; the concentration of these particles suggests failure of bearings. Chromates are used in cooling systems and their presence in lubricants causes the failure of coatings and gaskets. Silicon is mainly found as a result of silica dust introduced through air in the system. Sometimes silver is used for plating the bearings and as a soldering material; its presence in the lubricating oil indicates wear of such parts.

The concentration level of metallic elements depends upon their origin, and their permissible limit has to be decided depending on their nature and the type of equipment used. Sometimes copper and iron concentrations of several hundred parts per million (PPM) by weight have been found without any undue effect on satisfactory working of the machine and sometimes even less than 50 PPM concentrations have caused failures for some of the materials.

Contaminant monitoring techniques

To examine the presence of foreign materials in lubricating oils some other techniques can also be used. These may include spectrometric oil analysis procedure (SOAP) and the magnetic plug inspection system, besides ferrograph and X-ray spectrometry.

In ferrography, contaminants of size 0.1 to 500 micron range can be detected. The sophisticated instruments like Direct Reading (DR) Ferrograph can give quantitative measurement of wear particles. This instrument classifies metallic wear particles as either large (>20 micron) or small (<20 micron) and calculates the large to small wear concentrations. These values are entered into a PC-based Ferro trend database, which compiles the data points for wear trending analysis. A ferrograph is prepared in analytical ferrography where a non-wetting bassier is painted on one surface of a microscope slide.

This slide is kept in a strong magnetic field and the contaminated oil is allowed to flow over it. The particles get deposited on the slide depending on their size and magnetic behaviour.

Electron spectrometry can be used to examine samples of wear taken from the bearings of dissimilar materials. For the investigation of the chemical nature of ferromagnetic products, electron spin resonance spectrometry is used. By the use of ferrography technique, the different types of metal particles can be separated as per the particle size. The ratio of small size to big size particles may indicate the type and extent of wear.

Suitable size magnetic plugs are fitted into the oil-washed parts of a lubricating system to collect ferrous and other debris. Similarly, the wear-generated debris in the transmission and gearboxes can be captured using magnetic chip collectors fitted to the drain plug location of the gearboxes or reservoirs. The chip collectors can be withdrawn without loss of the fluid. The collected debris is analyzed to assess the status of the system.

The shape and colour of the metal particles can help to know the extent of deterioration taking place. The SOAP techniques identify the location and degree of wear by using atom absorption spectrometry. The X-ray fluorescence is a useful technique, though expensive, to analyze the magnetic elements.

The metallic particles produced due to wear also degrade the carrier oil/fluid, which can be revealed through oil analysis techniques such as layer chromatography.

Emission spectrometry (Electric spark)

A high direct voltage (1500 V) excitation of the metallic impurities in the oil sample causes different types of radiations depending upon the quality of the oil which can be spectrally analyzed. This instrument consists of a narrow slit, a prism to separate the component wavelengths of the radiation after it has passed through the slit and a photoelectric system to detect and measure the spectral radiation. The essential components of the atomic absorption type spectrometer are mentioned hereinbelow:

- An energy source (a hollow cathode discharge lamp) to emit characteristic light.
- An energy source to transport and atomize the metallic elements (flame).
- Wavelength selector, to isolate appropriate wavelengths of light.
- An energy detector to convert light energy into electrical energy.
- An acetylene flame normally used to atomize the metallic elements in the sample.

Chemical Effects

Chemical effects are monitored by the measurement of certain traceable quantities of chemicals that are an indication of wear, leakage, and changes in lubricating properties. For the above purpose, spectrometric oil analysis procedure (SOAP) is an effective laboratory-based technique, which analyzes the wear particles (10 μm) held in suspension in oil and assesses their concentrations. The basic steps, required in the analysis, involve sampling, spectrometric analysis, data interpretation, and validation of results. This is based on a regular programme where the wear rate is established for each element associated with a particular failure and the amount of each respective element is monitored for any change. From the multiple-element wear trend patterns and experience, a diagnostic indication of failure is determined.

The spectrometric methods of analysis have been described above. For the purpose of analysis, standard analysis programmes are prepared by blending an appropriate weight of an air-soluble organo-metallic compound with oils used in various systems to provide a series of standards for the metallic concentration.

An example of spectrometric results of condition-monitoring for an engine is shown in Table 4.2.

Table 4.2   An example of spectrometric results of condition monitoring for an engine

Wear metals	Fine (ppm)	Coarse (ppm)	Additive/contaminants	Fine (ppm)	Coarse (ppm)
Iron	2036*	274*	Silicon	487*	96*
Chromium	112*	47*	Boron	43	18
Lead	5	8	Sodium	61	40
Copper	48*	12	Magnesium	917	
Tin	242*	123*	Calcium	1547	
Aluminium	93*	48*	Barium	1	
Nickel	3	1	Phosphorus	2130	
Silver	0	0	Zinc	4129	
Molybdenum	264	32	Vanadium	1	
Titanium	8	0			

*Data interpretation

The spectrometric analysis of the oil sample shows large amounts of fine and coarse Iron. Iron is most common of the wear metals and may indicate wear debris generation from different areas of the engine, such as the cylinder lining, connecting rods, piston ring and rocker arms as well as the push rod.

Continued ...

Data Acquisition

The first step is to obtain the necessary vibration data in a systematic way; the second step is to evaluate the data to identify the specific problem with the machine.

There are many ways that the vibration data can be obtained and displayed for detecting specific problems in rotating machinery. Some of the more common techniques are as follows:

◆ Amplitude versus frequency
◆ Amplitude versus time
◆ Amplitude versus frequency versus time
◆ Time wave form
◆ Amplitude and phase versus rpm
◆ Phase (relative motion) analysis
◆ Mode phase determination

Physical Measurements

Some failures can be detected through physical measurements of certain quantities and are called the *primary effects*. The primary effects are the conditions that are recognized through measurable changes in the performance of the machinery; for example, any significant change in power loss can be associated with an increase in friction in the system or with the leakage of hydraulic oil from the system. The type of condition-monitoring techniques used for these types of effects are dependent on the application, where each case must be separately assessed and analyzed.

The various measurable deteriorations which can be noticed in any machinery, include the loss of output because of higher consumption of fuel/electricity/oil, reduction in operating speeds, and increase in levels of noise and vibration, and so forth. These deteriorations lead to frequent failures of the equipment. The methods employed to detect the changes due to such deteriorations are as follows:

Continued ...

The chrome readings indicate wear of the cylinder lining (since the cylinder lining is chrome plated) and also that of the piston rings. Gray cast iron with relatively thick coatings of plated chromium up to 0.2 mm (0.008 in) is most commonly used for piston rings. The presence of large amounts of tin indicates wear of Babbitt lined bearings.

The connecting rod and the rocker arm bearings have Babbitt lined surfaces. These bearings are usually tri-metal sleeve type, which have a hard brass or bronze layer with steel backing where the thin layer of the Babbitt is applied. Therefore, the presence of tin and copper indicates the wear of bearings. Silicon is usually an indication of a contamination of the oil with airborn sand/dirt (silica). Magnesium, calcium, phosphorus and zinc are part of the oil additive formulation.

Lubricating oil consumption rate

The lubricating oil, which is either used with fuel or separately for lubrication of the moving parts, adheres to the cylinder walls of engines and is burnt during the process. Therefore, some consumption of lubricating oil is anticipated which is the part of the fuel consumption. An abnormal consumption of lubricating oil indicates that some of the oil is passing through the piston rings due to their wear. Fuel dilution, for example, produces a reduction in viscosity and therefore may increase the oil consumption rate.

Air pressure ratios

In the engines, where superchargers are used, the efficiency of the turbochargers can cumulatively deteriorate, with the passage of time and may change pressure ratios.

Temperature changes

Due to prolonged running of gas turbines, their blades get affected and may even fail before the scheduled time because of abnormal temperatures. Even in the case of internal combustion engines, the overheating of cylinders due to a faulty cooling system can cause seizure of the cylinders.

Vibration changes

Misfiring of any cylinder of multicylinder internal combustion engines may cause serious vibrations. The vibration torque in these cases becomes very high and may cause premature failure of the system. This may also result in power loss of the engines. Due to defects in the bearings, the vibration level may also increase.

Physical Effects

These effects highlight changes in the physical structure or appearances, i.e. surface discontinuities or cracks possibly occurring due to fatigue or wear. A wide range of techniques for monitoring the physical conditions of systems are available. These are described below:

Liquid dye penetrants

These are used for the detection of cracks and surface discontinuities in all metals, glazed ceramics, plastics and other non-ferrous metals. For this purpose, electrostatic fluorescent penetrants are also used.

Magnetic particle inspection

In this process magnetic particles either in powder form or suspended in a suitable liquid are used to detect the defects. Spray, pouring or immersion techniques are used depending upon the type of the component. Magnetic flaw-detection inks consisting of fine powder of black or red iron oxide suspended in liquid (paraffin) are also used to detect cracks.

Strippable magnetic film

The cracks and other defects of a magnetic material can be detected by the use of self-curing silicon rubber solution containing a suitable dispersion of iron oxide particles. In this process the rubber solution is poured on to the area to be inspected and a magnetic field is induced around the area by a permanent electromagnet. Under the influence of the magnetic field, the iron particles migrate to cracks, which can be seen on the cured rubber as intense black lines. This technique is particularly useful in the areas where the visual access is limited.

Ultrasonic

The basis of ultrasonic testing is excitations at a high frequency of 20 kHz and above, beyond the hearing zone of a normal ear. Generating and using ultrasonic waves can detect small cracks since the wavelength of the ultrasonic frequency used is approximately equal to the size of many cracks. The metals with elastic properties can transmit ultrasonic vibrations, which will be scattered due to defects. The techniques available for ultrasonic testing include pulse echo, transmission, resonance and frequency modulation technique.

Pulse echo technique. A piezoelectric crystal is normally used to generate the ultrasonic waves. Three main types of waves generated are: (1) longitudinal, (2) transverse, and (3) surface waves. When crystals are applied parallel to the surface, longitudinal waves are generated. The basic technique applied in the case of pulse-echo, changes with the nature of flaws in the component.

Transmission technique. The continuous waves produced by a transducer are passed into the test component. Flaws reduce the amount of energy reaching the receiver and thus their presence can be detected.

Resonance technique. In this technique, the transmitter is moved over the surface of the object to be tested, and the transmitted signals are observed. Resonance in the presence of flaws will keep the transmitted signal low, as flaws cause the signal to become weak or disappear. This technique is particularly suitable for thin surfaces.

Frequency modulation technique. The frequencies generated from a signal transducer are continuously changed and the echo returns with a changed frequency if the flaws are present. By measuring the phase difference between the transmitted and received frequencies, the location of the flaws can be detected and measured.

Visual testing

It is seen that experienced inspectors can visually detect many defects such as surface cracks, weld defects, and misalignments. The visual inspection equipment includes mirrors, lenses, microscopes, enlarging projectors, comparators, photoelectric systems, fibre-optic scanners and closed-circuit televisions. A few instruments used for visual inspection are given below:

Borescope. For inspecting narrow tubes, bores or chambers, the borescopes are used. These are precision-built optical systems with arrangements of prisms, achromatic and plane lenses. The instrument is used for inspecting combustion chambers, fuel nozzles, compressors, and turbines, etc.

Fibre-optic scanners. These are of two types:

1. Coherent devices (image carrying) used for medical and military purposes, and also for quality control inspection.
2. Non-coherent devices (light piping) that are used for illumination purposes, and are available in the form of either glass or polymer materials.

Cold light rigid probes. These instruments are particularly suitable for combustible and heat sensitive areas, as they work on the image transfer principle and transmit cold light.

Deep-probe endoscopes. Working on the principles of fibre optics, these are used for the inspection of the pipe work in boilers and heat exchangers.

Pan-view fibrescope. Any defects in the wall of the tubes or in the internal parts of a machinery can be inspected by using this type of fibrescope. It can be inserted in the tube or internal part of a machinery providing forward viewing to check the defects.

Electron fractography. This technique is used to find out the growth of fatigue cracks developing in a structure as each fracture has its own fingerprints, which can be visualized by an electron microscope.

Temperature Effects

The variation in temperature of certain components of machinery can be used to predict failures attributable to corrosion, wear, fatigue, leakage, and poor design. This principle is successfully applied to power transmission lines, hydraulics, tyre defects, transformers, bearings, and electrical switchgears. For this purpose, contact sensors in the form of liquid expansion sensors (mercury or alcohol), bimetallic expansion sensors, thermocouples, thermocouple sensors, and resistance sensors, are generally used.

In a simple method of surface temperature monitoring, the use of paints, crayons or papers that change colour at known temperatures can be made use of. In a few cases, non-contact sensors based on radiation of energy can also be made use of. The various types of defects that can be detected through temperature monitoring are enumerated below:

Bearing damage

It is seen that due to surface damage to roller bearings on account of rubbing will cause an increase in heat generation if the bearing does not have a thermostatically controlled cooling system. This increase in heat generation will cause a rise in temperature at the bearing surface, which can be detected with the help of surface mounted sensors.

Effectiveness of lubricant

The excessive rise in temperature at the surface of a machinery would indicate either the failure of the coolant or the failure of the lubricant used. The rise in temperature can also be due to pump failure, pipeline blockage, or damage to the heat exchanger.

Heat generation variations

The uneven distribution of temperature on the casing surfaces indicates incorrect combustion in internal combustion engines or boilers. Locating the thermocouple sensors around the casing of the machinery and continuously scanning the output can provide data on the temperature distribution for remedial purposes.

Deposits of foreign materials

The blockage of pipelines due to sediments, ash, dust, corrosion, by-products, etc. can be detected by monitoring temperature at appropriate places at specified time intervals.

Damage to insulating materials

Damage to insulating materials can be easily detected by using a scanning infrared camera. Cracks and damages will be shown as hot or cold spots.

Corrosion Effects

The following techniques are available for identifying and determining the rate of corrosion for various applications.

Linear polarization resistance method (Corrator)

This technique is used for finding out the rate of corrosion in electrically conductive corrosion fluids. The concept here is to apply a 10 mV potential perturbation and measure the resultant current caused by this perturbation. The magnitude of the current signal is directly proportional to the corrosion rate. It is a normal practice to alternate between positive and negative perturbations so that the difference in corrosion rates measured in each direction of polarization can be examined. The corrosion rate is calculated from the polarization current. Either two or three electrode probes are used. This is a very fast and accurate technique that can be completed within 30 minutes.

Corrosion potential measurement

This technique is useful where the material shows both the active and passive corrosion behaviour in the process, and can be used to indicate the development of active corrosion. This can also be coupled with polarization resistance measurements as backup for the confirmation of high corrosion rate or may be used as the primary indicator for active corrosion.

Weight loss measurement

This is the most common corrosion rate measuring technique for general and localized corrosion. The weight loss of the material of known surface area is determined in the specific environment for a known time period. From this, the corrosion rate can be calculated although it is more time consuming compared to other processes.

It is noticed that much of the information enumerated above concerning condition-monitoring techniques has a few applications in the development of on-board machine fault detection systems. Many of these techniques require large capital investment and, therefore, will be more useful to the industries with high risk in terms of safety and/ or extensive loss of production.

Frequency of Condition Monitoring

The frequency of examining the condition of a machine or its components depends on a number of factors such as:

- ◆ Criticality
- ◆ Availability of a stand-by units
- ◆ Standardization of items
- ◆ Operating conditions
- ◆ Failure statistics (MTBF, MTTR)
- ◆ Cost of monitoring
- ◆ Cost of failure
- ◆ Cost of maintenance

We can use the following formula for determining the optimum examination frequency in the case of constant failure rate.

$$e^{\lambda t} - \lambda t = 1 + \lambda \left(\frac{C_1'}{C_q} \right)$$

where

λ = failure rate

t = examination interval

C_1' = examination cost

C_q = unitary downtime cost, i.e. the sum total of maintenance cost and the cost of production loss.

FUTURE OF CONDITION MONITORING

To extract maximum profit from the minimum investment in plant and equipment, the use of condition monitoring can be made to achieve these objectives in the following ways.

◆ By improving equipment reliability through effective prediction of failures with corrective measures
◆ By minimizing the downtime of equipment with better maintenance planning and scheduling
◆ By maximizing component life through better design and upkeep
◆ By utilizing condition-monitoring methods to improve and sustain equipment performance throughout its life
◆ By minimizing the cost of condition monitoring
◆ By developing an effective spare parts management system

Through better maintenance practices the performance of equipment/system can be improved. The level of maintenance can start from innocence level and can reach to the excellence level to deliver the desired results. However, the following areas will bring appreciable changes in respect of maintenance function.

Prediction of equipment failures

By introducing the maintenance functions at each stage of equipment working it would be possible to minimize the failure. Condition monitoring can be used to predict unplanned failures with its merits that the same is not used at regular intervals as in case of preventive maintenance. The important aspect of condition monitoring is to know the frequency of failure or the 'lead time to failure'. As on date no standard methods are available to predict the interval of monitoring, and in practice components may fail though they may be under the regular condition monitoring. Therefore, efforts must be directed to accurately find out the interval since the success of condition monitoring programme solely depends on it. No organization will be prepared to invest more without successful results.

Prediction of equipment condition

It is important to keep the holistic view of the condition of the equipment/system in any organization. During regular condition monitoring if it is noticed that the condition of the pump bearing is bad and need attention, it will be preferable to assess the condition of all other components of the pump in order to determine whether any of the other items should also be replaced or readjusted at the same time to save or minimize the downtime of the machine.

The integration of above activities is difficult because their data recording is done at different places. Integrated condition monitoring softwares are available in the market, which permit integration of oil analysis, vibration analysis, and other condition-monitoring data.

Accuracy in failure prediction

The calculation of reliability of a system or of a component call for accurate information as then only the results will be acceptable. In an automatic monitoring system, integration of data should be done properly. Sometimes the value of a parameter may change due to load conditions and therefore may not give accurate fault detection and failure prediction. To achieve better integration the systems must be able to interface with one another using common standard protocols. In some cases such built-up softwares are available in the market for condition monitoring. Some of the companies are working on this to maximize the total system performance. Therefore 'smart sensors' that permit on-board signal processing and analysis must be used in condition monitoring, though it is costly. In due course of time the smart sensors technology will greatly reduce the complexity of linking the outputs of these sensors to current process control systems and continuous on-line monitoring will become possible.

Cost of condition monitoring

As discussed above the use of on-board smart sensors must be both accurate and reliable in assessing equipment condition and prediction of machine/system failures. In the early days the accuracy of diagnosis of equipment was largely dependent on the skill and experience of the individual analysts. Now this dependence has been reduced due to development of more effective vibration analysis software, and the use of expert systems is being made too. If the smart sensor based technology has to work, it must be made cost effective. This will be possible with proper monitoring and feedback system employed for this purpose.

Reliability improvement

With the introduction of planned preventive maintenance, the next objective would be to minimize mean time between failures (MTBF). To improve the overall reliability of the machinery the weak components/sub-systems must be replaced at regular intervals during scheduled inspections. In this regard condition monitoring can help in many ways. Through vibration analysis the conditions of the bearings can be known and their expected life can be predicted in advance. The higher the overall vibration level the shorter would be the expected life of the bearings. The other things, which can be monitored, include misalignment, improper bearing installation, rotor imbalance, pump cavitations etc. However, the cost trade-off between more frequent and more rigorous monitoring and improved component or equipment life must be done.

Equipment performance optimization

In case of sophisticated equipment/systems such as steam turbines, pumps, diesel engines, etc. the performance monitoring will help not only the replacement policy of equipment but also the optimum and economic working of the system. This can be based on using measurements of temperature, pressure, power output, etc. in order to determine the equipment condition.

The summary of the above discussion is as follows. These measures can bring about appreciable changes in the maintenance function.

♦ The development of smart sensors and other online monitoring devices for cost-effective condition monitoring.
♦ Incorporation of in-built vibration sensors.
♦ Development of condition monitoring software with expert systems
♦ Interaction and integration of condition-monitoring data
♦ Use of condition-monitoring techniques to improve reliability
♦ Reduction in cost of condition monitoring

Knowledge Requirement

In case of an unmanned machining system, human supervision has to be replaced with a sensor called Real Time Actual Condition Knowledge (RTACK) data which can be used with an intelligent adaptive control system to optimize the machine operation in total and not simply to achieve sub-optimal goals. At present there are no sensors available to meet such requirements during machining operations. In some areas, non-contact optical sensors are being used to measure surface finish. For in-process alignment etc., vision or fast three-dimensional laser scanning systems appear to offer an effective solution if the same can be developed. To achieve the desired results, RTACK can be effective in the following situations:

- Multidimensional sensors in place of individual sensors for the control of system.
- Grouped sensors either in series or parallel configuration for the control of information.
- Multi-layered sensor input at various levels in a control system for optimum performance.

Management Process

For the control of various operations in an automated system including maintenance, the management of all such activities is a challenging job and therefore, must be given due attention. For this purpose, some software packages are available but a lot of work has to be done to overcome the control problems. However, as the computing and sensor power increases, the available techniques may be viable through intelligent machining systems. To form a smart machining system, an integrated approach that can incorporate adaptable control with use of condition monitoring may yield the desired results. An online data acquisition system (DAS) can also be used to record and analyze the data for the purpose.

System Approach to Condition Monitoring

Monitoring of individual components/parts involves recording of information for result-oriented operational control, which is not efficient. Here a separate monitoring sensor needs to be used independently to measure each characteristic, for example, bearings of a shaft. When the concept of whole system is employed, the output of one sensor can give information on more than one sub-assembly. Therefore, monitoring at sub-assembly level reduces the number of sensors, which is the basic concept of system condition monitoring technique. The classic systems use the Laplace transform represent-ation of the differential equations describing the dynamic system under consideration. Practical condition monitoring does not need the analytical tools used in dynamic analysis but the concepts of systems approach of input/output relationship is being used.

Condition monitoring diagnosis is essential and important to study the state and condition of the equipment which can be done in either mode—routine or continuous. Depending upon the signal received, the failure can be categorized as hard or soft. In case of hard failure, many changes are observed in data, whereas in case of soft failure, very few changes are noticed. Signature matching can make this comparison, which is a common practice in vibration analysis.

CASE STUDY

Unbalance of double squirrel cage fan

The fan of a double squirrel cage induction motor experienced vibrations and associated noise in a mine. The supplier changed the bearings and adjusted the belt but with no success. At a later stage when a thorough investigation was carried out with the help of a vibration analyzer, it was found out that the problem was associated with fan imbalance. What followed was a successful single plane balance with measurements from bearings. Weights were added at the inboard end of the overhung fan wheel on its backing plate close to the bearings. Subsequently, there was a substantial drop in vibration level at the bearings. The problem remaining was attributed to:

♦ Inadequate axial support
♦ The variable pitch motor sheave
♦ Sheave misalignment.

The noise level also fell with the reduction in vibration level. The recommendations made were as follows:

♦ Replace the motor's variable pitch sheave with a fixed diameter sheave of the proper size.
♦ Align and tension the belts to the required specifications.

QUESTIONS

1. How does condition monitoring influence the maintenance activity function? Explain.

2. Explain the various levels of condition monitoring.

3. Describe the condition-monitoring techniques used in maintenance practice.

4. How does the information on dynamic effects help in assessing the condition of an equipment/system?

5. Explain particle contamination effect and its effects on equipment health.

6. Differentiate between the spectrometric oil analysis procedure and the magnetic plug inspection system.

7. Under what conditions is emission spectrometry employed and why? Explain briefly.

8. How can the chemical effects be useful in assessing the condition of an equipment? Explain briefly.

9. List the physical parameters which can monitor the condition of an equipment/system. Discuss in detail.

10. Describe the monitoring techniques used for ascertaining the physical condition of an equipment.

11. Briefly discuss the instruments used for visual testing.

12. Which components/parts can be monitored using the temperature effects?

13. Briefly explain the techniques which can be used for the detection of corrosion in a machinery.

14. Enumerate the cost of components for a condition-monitoring system. Under what circumstances will condition monitoring be cost effective?

15. Develop a list of commercially available condition-monitoring instruments. Comment on their applicability.

16. Discuss the factors that must be considered for switching from PPM system to a condition-based maintenance system.

17. Discuss the future scope and limitation of condition monitoring.

Chapter

5

Maintenance Planning and Scheduling

INTRODUCTION

Planning is the activity performed to make a time-bound programme for successful completion of any job. To correctly and efficiently perform the planning function, management should provide adequate guidance on the levels of control necessary to ensure consistent quality maintenance of plant/equipment. The basic concepts of planning are applied to the maintenance function as well. Maintenance planning and scheduling should be viewed as the centre of industrial maintenance since other processes such as preventive maintenance, root cause analysis (RCA), spare parts management are dependent on the planning and scheduling processes to work.

All three approaches to planning, namely long-range planning, short-range planning and planning for immediate activity are also applicable to maintenance work. For effective working of any system it is necessary to plan and schedule its activities. The planning approach differs widely depending on the purpose and the work being planned. A long-range planning is done at the higher level. In this approach of planning the goals are set, strategies are developed and operational programmes are devised for a period of five to ten years. Short-range plans are prepared at the departmental level for a period of one to two years. Immediate activity planning is almost a routine procedure and is done at the working level as and when required. The three approaches to planning have very little in common except that they all come under the area of planning and all are necessary to fulfil the objectives of the organization. In any planning, the basic steps required are, problem identification, identification of solution, evaluation, selection, and placing of the solution into practice. Maintenance planning also involves a similar decision-making process. The basic objective of any planning, therefore, is to convert the concepts into workable actions. In the maintenance context, planning is therefore the task of organizing resources to carry out a job satisfactorily at reasonable cost within a specified period of time.

PLANNING OF MAINTENANCE FUNCTION

Maintenance planning involves the assignment of jobs to the maintenance crew with relevant information of the work to be carried out. Job assignment must be done on the basis of proper job scheduling of the maintenance work. The risk of an emergency maintenance work going unattended must be minimized. At the same time the maintenance workforce must not remain unutilized. The following aspects need to be considered carefully while planning the maintenance work.

Job Distribution

In planning and organizing the maintenance work, the first and foremost consideration should be given to the distribution of the jobs to the personnel for preventive as well as emergency maintenance work. If the same group is used for both the functions, some amount of time must be allowed during scheduling for unforeseen breakdowns, which may occur, to avoid underutilization or over-utilization of the available manpower. If separate teams are used for preventive and emergency maintenance, the effective utilization of manpower calls for proper planning of maintenance functions. Such a planning process should incorporate the following components.

Organizational goals

The maintenance planning must be in accordance with the organizational goals, but all aspects of the organization do not necessarily form part of maintenance. Maintenance Needs Analysis (MNA) should therefore be carried out to identify the role of a piece of equipment or asset in achieving the organizational goals. Based on such an analysis, equipment or assets are identified for inclusion in the planned maintenance system. The objectives of maintenance planning at the departmental level must be clear to all the personnel involved in the work and they must also be clear about the objectives of the organization. The planned work should be carried out to improve the performance of the production system.

Policies

The maintenance department must evolve the fault-oriented maintenance policies. Such policies should identify the level of maintenance required for different assets and also formulate the procedures of conducting the maintenance operations. The policies must be well defined so that the maintenance functions planned can be easily identified for timely implementation. Here also the objectives of the organization must be kept in mind for efficient working of the maintenance functions.

Procedures

The maintenance planning involves determining the methods to carry out each of the maintenance functions. The procedures of the work must be well defined and standardized so that the maintenance personnel do not get into any confusion. The development of procedures includes the design of the inspection form, interval of reporting, issue of maintenance tools, etc. The procedure for organizing men, materials or other resources for carrying out any maintenance work should be well known to the maintenance workforce. This reduces the logistic time during any maintenance operation, without impairing the effective working hours of a machine.

Programme

The objective of any planned preventive maintenance programme is to reduce the system costs for providing services. A maintenance programme is a well-formulated combination of the available skills and resources that ensures optimum utilization to successfully complete the work. The development of such a programme involves (i) selecting activities for preventive maintenance, (ii) determining their frequency, and (iii) deciding which out of the two, whether repair or replacement, is more cost effective. These three aspects of a maintenance programme are briefly discussed below.

Selection of activities

To select the activities for programme implementation, the following information should be available in respect of each acitvity.

  (a) Frequency of failures
  (b) Causes of failures
  (c) Cost of failures
  (d) Cost of preventive maintenance
  (e) Impact of failures on the system

The selection of any maintenance activity is dependent upon the cost of maintenance and the importance attached to it. If the cost of the preventive maintenance is lower than the cost of the breakdown maintenance, then the former will be preferred over the latter. However, the requirements of the equipment/system must be kept in mind before the final decision.

Determination of preventive maintenance frequency

The optimum period or frequency for preventive maintenance can be determined from the failure data. The time period for different types of breakdowns to occur can also be calculated by analyzing such failure information. From this information, the interval of preventive maintenance is determined.

For the efficient planning of preventive maintenance work it is essential to know the cause of actual failures so that such failures can be prevented in future. It is known that a substandard material or workmanship causes frequent failures. This aspect is duly taken care of during planned/scheduled maintenance work.

The comparison of the preventive maintenance cost with the breakdown cost should also be made before selecting a particular policy.

Repair versus replacement

In the case of failures where no repair is possible such as bulbs, electrical goods, electronic goods, the defective parts have to be replaced. Such replacements are carried out during the scheduled inspections as well as on their failures. This type of scheduled replacement is similar to the preventive maintenance programme. Policies followed with regard to scheduled replacement are:

(a) Replace all items (both good and failed components/units) periodically.
(b) Replace only the failed components/units periodically.
(c) Replace the failed components/units as and when required.
(d) Replace the failed components/units as they fail and also replace both good and bad components/units periodically.

The decision regarding the scheduled replacement policy has to be based on the analysis of the past failures data on the same type of projects or products from the same field. Incorrect decisions may, in fact, increase the maintenance cost to a high level.

MANPOWER ALLOCATION

The most important task of the maintenance management group is to provide adequate manpower for various jobs. This also should take care of the skill levels of the personnel involved. This is not an easy task since many variables enter into the calculation of manpower requirements. In the absence of proper maintenance records and planning, it is impossible to establish the proper utilization of the maintenance workforce, which in turn may lead to criticism by the authorities for underutilization or unnecessary overtime payments.

Though measurement techniques may have been employed to arrive at the total work content of activities, unnecessary work associated with such activities must be avoided. This unnecessary work can be minimized if planned maintenance methods are adopted.

In addition to planned maintenance, machine modifications, project re-arrangement, etc. also require the services of maintenance personnel. The manpower required for such work will depend upon the nature of the work and should be taken care of during manpower calculations.

Manning levels can therefore only be established after proper planning and control over maintenance and by keeping reliable records of analysis of the results of maintenance.

Sometimes trained persons are not readily available for the maintenance work and then it becomes necessary for the maintenance department to train the people through formal technical education and on-the-job experience.

There is a general feeling among the management personnel that a person classified as maintenance engineer, can undertake all types of jobs. This is not true. Each job of maintenance requires the placement of an individual who has the true ability and the right attitude for the specified job. The estimation of maintenance manpower requirements becomes easy when maintenance types, failure frequencies and the number of equipment involved are known in advance.

Staffing Maintenance Function

Staffing is dependent upon the ability of the organization to handle both preventive and emergency maintenance requirements at the same time. The advantage of the preventive maintenance is that the work can be planned and scheduled properly for the effective use of available manpower or the manpower can be increased as and when required. To make a proper assessment of the manpower requirements, the available techniques such as Gantt chart, material analysis, sequencing and linear programming can be used.

The emergency maintenance staffing presents a more difficult problem to the management due to random nature of failures. It takes time to even assess the total work for finalizing the manpower requirements.

It is necessary in this case to provide the required manpower in such a way that the maintenance function can be achieved at optimum cost. The past information on a specific case will help the planner to use the analytical methods available for determining the required manpower.

After staffing a maintenance function, it is essential to assess its effectiveness. This can be done by the following methods.

Work-time versus *idle-time*

For the calculation of the effectiveness of maintenance personnel, the work-time and the idle-time of each worker should be recorded in a way which can accurately reflect whether a worker has been utilized to the given potential or not. Remedial actions can also be initiated, if required.

Control of labour performance

For a given period of time, the labour efficiency can be calculated to reflect the effective utilization of the workers. This can be determined as:

$$\text{Labour efficiency} = \frac{\text{Standard hours earned}}{\text{Hours worked against standards}}$$

It may be observed from Figure 5.1 that the efficiency of a labour improves with a reduction in working hours, however, some minimum working hours need to be assigned to each worker.

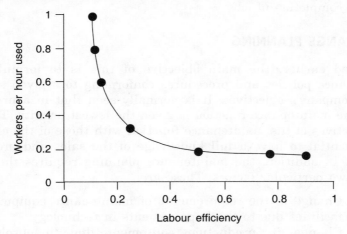

Figure 5.1 Labour efficiency.

Utilization of equipment

It is evident that lots of service equipment/tools are used for the maintenance of equipment. Their utilization factor can also serve as an indicator of the effectiveness of maintenance personnel. On the other hand, maintenance work also suffers sometimes due to non-availability of the service equipment and tools in time. This aspect should be given proper attention for effective working of the maintenance personnel.

Effect of Planning on Manpower Utilization

If most of the jobs related to maintenance are planned effectively, the utilization of manpower also improves. It is seen through the studies that there is 15–20 per cent saving in manpower with the planned maintenance function. The downtime of the equipment is also reduced through maintenance planning. The following additional benefits also accrue from planning.

1. A well-planned work progresses without interruptions and is sometimes completed even before its scheduled time period.
2. The instructions for the jobs are available in a clear and well-defined manner.
3. There are no problems regarding the availability of spare parts as the same are ordered in time.
4. The tools/special equipment required can also be arranged before the start of the maintenance work.
5. Supervision of the works is also better because the foreman concerned also gets sufficient time for every job under his control, which further helps in overcoming delays in the completion of jobs.

LONG-RANGE PLANNING

As stated earlier, the main objective of this is to formulate the maintenance policies and procedures conforming to a given standard of the company's objectives. It is normally seen that in a production plant, the maintenance function is given the lowest priority. To match the objectives of the maintenance function with those of the company, it is essential to have a full knowledge of the sales and production forecast. In addition, the maintenance planning requires the projection of two particular factors. These are:

1. Change in the requirements of maintenance equipment and facilities due to the improvements in technology.
2. Change in production equipment due to obsolescence, increased mechanization, automation, higher machine output and other technological advancements.

It is, therefore, necessary for the maintenance department to incorporate the above changes in their long-range planning.

The long-range planning of the maintenance function must be carefully coordinated with the planning of the organization. This may eventually help in reducing the maintenance costs.

The other aspect of long-range planning is concerned with the existing facilities. The plans must be flexible and adaptable to changes, which may occur in due course of time. A new type of equipment may pose serious maintenance problems for which the manufacturer's help should be taken. At times it is seen that companies do not plan for emergencies and thus incur heavy losses. Sometimes even public relations do affect the maintenance work, and this aspect therefore needs special attention during the planning stage.

DEVELOPMENT OF MAINTENANCE DEPARTMENT

The needs of the maintenance department in the next ten to fifteen years must be considered and kept in mind. Equipment frequently become obsolete due to advancements in technology and this should be taken care of by the department without incurring high costs. It may be difficult to get trained workers for the future maintenance activities, therefore, apprentice programmes need to be initiated at the right time. Maintenance personnel must be trained for maintenance work, along with the training for operating personnel, during the procurement of a new equipment. Some training programmes must also be planned for senior employees for updating their knowledge of the latest techniques available. The organizational structure of the maintenance department should be such that the work does not suffer on account of the new training requirements imposed on it. There can be any of the following structures for the maintenance department.

Decentralized. In this type of structure the maintenance work may be done as per the requirements and no specific structure for the maintenance function need be established. Normally all the persons in the maintenance department are capable of doing the relevant jobs. No specialist is required for a particular job.

Centralized. In this case, for each job a specific shop is established and the workers therein are the specialists of their own jobs. For example, an engine shop will deal with all types of engines only, or a motor winding shop will deal with the motor winding work only.

During the development of the maintenance department, the following points must be considered.

Coordination

Efficient coordination among workers can be achieved by following the common guidelines and clearly specified lines of authority. A good leadership can bring people together and can create congenial atmosphere for working. This is important at all levels of the maintenance department.

Development of subordinates

The development of subordinates depends upon the supervision and guidance provided by the supervisors. Since man is an egotistic and emotional creature, special care should be taken to deal with people particularly in low profile jobs such as maintenance. During assignment of jobs, an individual's skill, education, intelligence, temperament and motivation, etc. must be kept in mind. This will help in developing persons for the required jobs almost as per their choice.

Proper utilization of manpower

The organization must be in a position to utilize the capabilities of each individual to the maximum extent possible. It is seen that a particular person may not be a good supervisor but an excellent engineer/executive. Therefore, the organizational structure should be such that the effective utilization of each individual is possible in the interest of the company.

Conflicts

As in any other system, conflicts may occur in the maintenance department too. The duty of the departmental head is to resolve the conflicts in such a way that the work does not suffer. For this purpose, better coordination and integration within the system is desired. If the conflicts are not resolved in a congenial manner, the maintenance work will suffer in terms of quantity and quality.

SHORT-RANGE PLANNING

As discussed earlier, the maintenance work requires looking into day-to-day problems which may include random breakdowns. Planning for such jobs has to be made in advance, which can be termed short-range planning. Here an emphasis has to be on the nature of jobs that require routine check-ups and inspections as per schedules.

Short-range planning may concern installation of new equipment, cyclic works, and preventive maintenance work. These aspects are discussed below.

Installation of New Equipment

The maintenance department may face some problems during installation of new equipment and for this the following questions need to be answered.

(a) What is the additional number of equipment to be installed?
(b) Does the equipment design match with the installation requirements?
(c) Do the layouts conflict with the existing foundations?
(d) Do time schedules permit installation?
(e) Is sufficient manpower available for installation?
(f) Are special tools readily available for the installation purpose?
(g) Has the requirement of maintenance force been studied?

After the answers to all of these questions are obtained, planning can be done for the installation of new equipment. The above information will also help the planner to procure the facilities not available within the organization and the views of experts if required can be sought beforehand. This consolidated approach will also help to eventually save some time, which otherwise will be spent in gathering the required information in an ad-hoc manner.

Cyclic Works

Maintenance work of repetitive nature such as overhauls, painting, routine cleaning, etc. can be planned a few months in advance to minimize interference with other programmes such as new install-ations and facility expansion. This planning will help to undertake such works, normally during the shutdown periods of the equipment. This type of planning must get adequate consideration in the maintenance department in the interest of efficient operations.

Preventive Maintenance

During short-range planning, preventive maintenance works such as routine inspections, repairs, etc. can be included on monthly, quar-terly, and annual basis. This will help to schedule new installations, cyclic works and preventive maintenance work without disturbing the other functions.

PLANNING TECHNIQUES

Planning techniques are required to develop the overall master plan and for scheduling the major work of planning and execution. The Gantt charts and the milestone methods have been used in the past, but they have certain limitations in both planning and control. The use of Critical Path Method (CPM), Programme Evaluation Review Techniques (PERT), Least Cost Estimating, Scheduling Control and Automation by network methods have become more effective in engineering applications. Some commonly used techniques are briefly mentioned below. Any standard book on industrial engineering may be referred to for a detailed description of these techniques.

Gantt Chart

Henry L. Gantt developed a technique, called the Gantt chart, during World War I, which is widely used even today for planning. Individual jobs are described on the left-hand side portion of the chart, scheduled times are plotted on the right-hand side on the horizontal scale, the length indicating the job duration is shown as an open bar. A solid bar indicates the actual time. By looking at the chart, the status of a particular job can be ascertained, i.e. whether it is ahead or behind the scheduled time period. This chart, however, cannot indicate the delays occurring in the overall project period.

Milestone Method

During World War II, this method was developed for comprehensive planning and control. It is a refinement of the Gantt chart where an individual block or milestone is indicated within the open horizontal bars, each milestone representing a defined point in time. Though it is an improved version of the Gantt chart, it still lacks any predictive quality.

Critical Path Method (CPM)

Originally this method was developed to solve the scheduling problems in the industrial organizations. It does not make use of probabilistic job times but is of deterministic nature and considers variations in job times. Here more emphasis is given to the critical jobs so that the overall time of the project can be minimized.

This method argues that the duration of a project can be reduced if extra resources (men, machines and money) are provided for the jobs. A concept of crash and slack is applied to each job and an emphasis is placed on crashing all the jobs to reduce their periods or times and complete the tasks at the earliest. If all the jobs cannot be crashed, then only the critical jobs should be crashed to expedite the maintenance work. The critical path method assists in identifying the jobs to be crashed. The time–cost trade-off relationship for a typical job is shown in Figure 5.2.

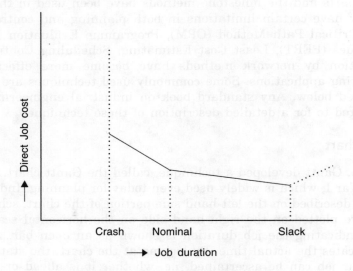

Figure 5.2 Time–cost relationship.

Program Evaluation and Review Technique (PERT)

The use of PERT is very common and can be applied to all the industrial projects. This takes the uncertainties into account and assumes three types of time estimates, i.e. optimistic times, pessimistic times, and probable activity times. Based on these, the average or the expected time of an activity can be calculated by the formula given below:

$$t_e = \frac{t_o + 4t_m + t_p}{6}$$

where

t_e = expected time
t_o = optimistic time
t_p = pessimistic time
t_m = most probable activity time

The PERT technique can very well be applied to maintenance work where uncertainties very much prevail. If the exact times of the activities are known through time and motion studies, then the CPM/PERT network can be used to find the initial time for the completion of the job, which helps the planner to focus attention on critical jobs as well.

PLANNING PROCEDURE

The most important aspect of maintenance planning is its procedure. It can be a four-step process as mentioned below:

1. Organizing maintenance resources to ensure their effective use in future
2. Scheduling the resources for the planned period
3. Execution of plans according to the schedules
4. Establishing a feedback system for all the above processes to know the deficiencies of each of the processes.

The maintenance planning should also evolve a detailed organizational chart for the maintenance department including assignment of responsibility and a detailed procedure for feedback and control mechanisms.

Planning Inputs

Like any other planning job, maintenance planning needs acquisition of certain information as the planning input. These inputs include:

♦ Nature of the equipment to be maintained and their number
♦ Resources required for the maintenance jobs
♦ Maintenance activities and time required for these activities

- ◆ Information on availability of skilled manpower
- ◆ Information on hiring and contract services
- ◆ Information on maintenance tools
- ◆ Maintenance instruction manuals
- ◆ A possible maintenance control policy

After collection of these planning inputs, maintenance planning is carried out to assign the following:

- ◆ Target period for job completion
- ◆ Priorities of jobs
- ◆ Job scheduling by preparing the Gantt chart, or PERT, or CPM network, etc.

ESTIMATION OF MAINTENANCE WORK

During planning, it is important to estimate the quantity and quality of the maintenance work required. This will help in allocation of the adequate manpower and material in time. In practice, it is observed that the maintenance work suffers because proper estimation of it is not done in time. The following three methods can be used for the estimation of maintenance work.

Measurement by Estimates

This system is more useful where the material cost is negligible compared to the labour cost. It therefore consists of comparing the actual cost of labour with the estimated costs. The benefits of this method can be listed as follows:

(a) The estimate of the total time can be prepared before the start of the job.

(b) The estimate provides the basis for scheduling the maintenance work.

(c) The estimate also provides comparison with the actual manhours charged during a particular operation.

Measurement by Historical Data

It consists of statistical analysis and treatment of past time data on completed jobs to arrive at an average, which can be considered a suitable standard. This method compares the actual manhours required for the job with the standard index based on average historical time for the same type of job. This method also provides a basis for the estimation of a new type of work, where details are not available. Some similarity among the two jobs can be of use for the estimation of maintenance work required. The advantages of this system include:

(a) Low cost of administration, because the required data can be collected easily
(b) Adaptability to routine and repetitive types of jobs
(c) Easy installation for estimation of maintenance work
(d) Provision of immediate but limited improvement in labour effectiveness.

Measurement by Conventional Standard Time Data

Under this method the work is measured with standard time data by breaking down a job into its basic components and applying time standards for each component. The sum of these standard times plus the various allowances for job conditions represents the time that should be allowed for the job. A comparison of this allowed time with the actual time measures the job effectiveness.

The work measurement, based on standard time data, can be accurate only if the application of the appropriate standards is done correctly. If the wrong data are applied or any element is omitted, the result will be incorrect. The advantage of this method is that sufficient time is allowed for each job and thus clustering of the various operations is avoided.

MAINTENANCE CONTROL

The economic aspects of maintenance work become more pertinent for developing countries as the capital equipment cannot be replaced fast. This necessitates an efficient maintenance system along with the well-equipped maintenance base and availability of indigenously developed spare parts. However, the most pertinent constraint of maintenance management is to maintain equipment at reasonable cost. This requires effective control of the maintenance budget.

In big organization, a separate budget is usually set aside for the maintenance work. It is the duty of the maintenance manager to effectively utilize this budget. For this purpose, the maintenance department may need scientific forecasting of maintenance needs so that the overall mission of the organization can be achieved. Budgeting for maintenance work needs identification of various cost elements. Such cost elements depend on the nature of the industry and the degree of mechanization and technological involvement. Broadly, such elements may be categorized under the headings such as *equipment, labour, external service,* and *overheads* of the department.

For the budgetary requirements, it is essential to estimate the economic life of an asset. Economic life can be calculated on the basis of the available data. However, this cannot be generalized as the working conditions vary depending on the type of industry, its location, and environmental conditions. The estimation of the economic life plays an important role in controlling the maintenance decisions concerning repairing, replacing, or modernizing a plant by an alternative technology. For this purpose, standard procedures need to be evolved such as techno-economic audit of system or equipment.

Maintenance control also involves integration of accountability within the system. Proper accounting of maintenance work should be carried out at every level of the maintenance organization. Pursuance of such accountability based control improves the quality of manpower and material utilization. It can also suggest the areas where the maintenance requirements need to be increased or where an assessment of the abnormal maintenance costs needs to be made. If such details are prepared equipment-wise, it helps in taking the right decision at the right time.

MAINTENANCE SCHEDULING

The concept of scheduling can be applied to the maintenance function for improving the operational availability of the equipment. To start with the maintenance scheduling, the first step would be to know the number of machines to be maintained. The next step would be, how to maintain them. Then the maintenance schedule is to be prepared for all critical items, which require preventive maintenance. Initially, this exercise can be started with a few selected components of a machine and with experience the knowledge gained can be applied to the complete system.

The main objective of maintenance planning and scheduling is to raise the standards of the maintenance functions and make it cost effective which is possible only through critical analysis of the results obtained. In the case where the number of machines is high and the manual scheduling is difficult, the use of computers is inevitable. For the jobs not completed in time, rescheduling of the same can be done.

Once the process of schedule generation is completed, all assigned works must progress smoothly until something comes up that necessitates an alteration in the predefined process. Sometimes due to the budgetary constraints the rescheduling of work become essential. Backlogging work must be minimized and avoided. Basically, the scheduling is of two types—fixed scheduling and dynamic scheduling.

In the case of fixed scheduling the problem of backlogging is common when the work is not completed as per the schedule and the system can work under ideal situation. However, dynamic scheduling takes care of all practical constraints under all circumstances.

For proper scheduling of the maintenance work, the job must be controlled through the work-order system, which provides the basic paperwork for planning the workforce, the material and the time. The work order is also an authorization to carry out the particular job. The concept of priorities is also essential in case of scheduling.

The priorities must be set to handle the mixture of backlogged and the new pieces of scheduled jobs. The frequency of maintenance and the criticality of the operations are the key factors for establishing the priorities for the jobs. Frequency indicates the relative importance of one procedure over the other. If the equipment becomes over due for a certain number of jobs, it is treated critically important and the priorities must be set accordingly. To minimize such situations the scheduling must be done when it is due. For effective results, under-scheduling and over-scheduling must be avoided.

Scheduling process: The scheduling process consists of the following activities:

- ◆ Prioritizing the maintenance work
- ◆ Dealing with the emergencies
- ◆ Maintenance calendars
- ◆ Scheduling the backlog
- ◆ The allocation of scheduling methods
- ◆ Priority numbering systems
- ◆ Weekly and daily schedules
- ◆ Controlling the backlog

Cost reduction with maintenance planning and scheduling

The cost reduction in maintenance function is possible with proper planning and scheduling of the work. The first step in this direction is to utilize the shutdown time of the equipment occurring due to failures. This will depend upon the following:

- ◆ What type of work is executed during shutdown
- ◆ When is the shutdown work list finalized
- ◆ How well is the shutdown work being planned

Work executed during breakdown: Attempts must be made to minimize the shutdown period of the machine, which ultimately will reduce the maintenance costs. During the shutdown period, jobs of minor nature such as cleaning, repair and preventive maintenance, etc. should be undertaken. When this basic principle is followed, it will result in reduced overtime, lower contractor costs and better documentation.

It is a common belief that during the major shutdown period maximum maintenance related works must be completed. There is also pressure to postpone the scheduled minor repairs by executing this work during the major outage instead. In such an approach, it is likely that additional cost attributable to overtime labour, expedited parts delivery, execution of unplanned work and reduced worker efficiency will exceed the apparent savings. Major shutdowns should therefore never be used for periodic minor repairs. Minor downtime is also important, because it provides an opportunity to do preventive maintenance and repairs that cannot be done during the operational time.

It is noticed that increasing the Mean Time Between Production Loss (MTBPL) versus Mean Product Loss (MPL) indicates that the efforts of maintenance and operations are resulting in improved productivity. Increasing the time between schedule outages or reducing the time available for repairs during scheduled outages may cause this trend to turn downwards due to increased unscheduled breakdowns.

Shutdown work finalization: In the long-term plan, there should be fairly detailed lists of major works that must be done during each shutdown in the coming years. For example, during boiler inspections, relining of the tile tanks, sewer repairs and electrical power distribution system inspections should be planned and estimated in the long-term plan. Some mandatory provision must be made for minor repairs as inspections often do not get adequate attention until it is too late to properly prepare for their execution.

In order to get a major shutdown accurately budgeted, the scope, the duration and the timing of the outage must be supplied before the budget is approved. The short-term plan can be developed using the long-term plan as a starting point. During the short-term plan, expensive repairs must be identified. When the equipment is nearing its final outage, then the short-term outage is not justified. Jobs added on short notice before a shutdown, are generally the cause of most disruptions to planned and scheduled work. It is important to note that planning the maintenance work is expensive and its cancellation at the last moment is wasteful. Review of repair histories and making of accurate estimates of the time, materials and expenses that commonly occur with each project is essential.

The success of shutdown planning and scheduling depends on key events occurring far in advance. Continuous improvement of the process requires that a detailed critique of each shutdown be performed, with inputs solicited from maintenance operation engineering and supply stream personnel. A word of caution. Never estimate the budget for shutdowns using the budget figures from the past.

Planning shutdown work: More work can be done by less people if it is properly planned. The resulting repair quality will be better with the reduced costs. The unplanned repair works which take 8 hours to complete will take 2 hours when planned. This is due to better instructions, and coordination of tools and materials and resources. Each work order should be planned before execution. This includes all the preventive maintenance work as well as repairs. The planning should include the following:

◆ A clear scope of work required
◆ An accurate estimate of the manpower required
◆ A detailed procedure for the work
◆ A list of tools and special equipment required
◆ Schedule of execution of work
◆ Safety and environmental hazard

The comparison of quality shutdown management versus the poor shutdown approach shows the areas where savings in maintenance functions are possible.

Quality shutdown	Poor shutdown
Controlled work list	Unlimited work list
Limited overtime	Unlimited overtime
Routine work done in scheduled time	Routine work done in overtime
Timely availability of spare parts	Untimely availability of spare parts
All work planned in advance	Some work planned, most unplanned
Accurate schedule	Unscheduled work
Add on work is rare	Add on work is common
Planned, scheduled work get priority	Planned scheduled jobs get cancelled
Budget based on reality	Budget based on past information
Better documentation	Poor documentation
Backlog minimum	Backlog maximum
Absenteeism of workers is less	Absenteeism of worker is more
Better acountability of work	Poor acountability of work

QUESTIONS

1. Describe the role of maintenance planning in the maintenance function.

2. How can the maintenance function be planned for effective working of the maintenance department?

3. What is the importance of manpower allocation in the context of maintenance function?

4. How does the staffing maintenance function help in improving the maintenance related activities?

5. What effect does planning have on manpower utilization in a maintenance organization?

6. How does the long-range planning help the maintenance function effectively? Explain.

7. For efficient discharge of maintenance work, which type of maintenance department can be appropriate and why? Discuss.

8. Explain the role of planning techniques in the efficient performance of the maintenance functions.

9. Describe the activities of the maintenance function, which fall in the category of short-range planning.

10. How does the estimation of maintenance work help a planner to improve the working of a maintenance organization?

11. How does the maintenance scheduling help the maintenance department? Explain briefly.

Chapter
6
Motivation in Maintenance

INTRODUCTION

Motivation is the result of the interaction between the individual and the situation. Individuals differ in their basic motivational drive since it is driven by the situation. So, motivation is the process that account for an individual's intensity, direction and persistence of effort towards attaining the goal. Intensity is concerned with how hard a person tries, however, higher intensity is not always likely to lead to favourable job performance, until the efforts are not channelled in proper direction that benefits the organizations. Therefore, quality of efforts is important as well as, its intensity. Persistence is a measure of how long a person can maintain his or her effort. The motivated personnel stay with a task long enough to achieve their goal as compared to other persons.

The main problem faced by maintenance managers is the control of the workforce and their motivation. The motivation function has direct bearing on the following:

◆ Job performance (ability)
◆ Productivity (skills)
◆ Job satisfaction
◆ Employee hiring and firing i.e. extension

The managers are forced to offset increasing labour costs by attempting to raise the levels of workers' productivity. In a number of Indian production systems, the availability of equipment is often poor. This is due to lack of adequate steps to improve the maintenance organization. One important cause of poor maintenance performance is the low motivation to work. Maintenance staff can be motivated through proper training, through providing incentives, through labour performance analysis and control, and through maintenance-based compensation. The concept of overtime to improve productivity has not yet yielded the desired results. Motivation means stimulation to act. Stimulation may differ from one type of job to another. The motivational efforts should be directed at stimulating men to work for achieving the desired goals of the company. However, the true motives of individuals are not always apparent and are best known to the individuals themselves. The actual motives of individuals may be to achieve immediate goals such as money, security, or prestige. Adequate importance should, therefore, be given to any type of job related to maintenance, in order to keep the morale of the maintenance workforce high. As seen in the past, the maintenance work is not given as much importance as the production/design work which causes de-motivation of the workers. Some theories to motivate the people in the maintenance department are as follows:

◆ Content theory (Need theories)
◆ Maslow's hierarchy of need
◆ McClalland's 3-Need theory
◆ Herzaberg's Dual factor theory

Motivation process: The Motivation process includes the following:

◆ Intrinsic reward (intangible).
◆ Extrinsic reward (tangible).
◆ Punishment is the process of administering an undesirable consequence for a desirable or undesirable behaviour from the workers.
◆ Extinction is the process where the principle of ignorance is applied for the undesired behaviour of the employee.

For the proper working of maintenance system the concept of goal setting theory can be applied; which lists the following:

◆ Matching of personal and organizational goals
◆ Rewarding goal achievement to the deserving candidates
◆ Self-management
◆ Goal acceptance

CHARACTERISTICS OF MOTIVATION

The following are the characteristics of motivation.

◆ It concerns with what activities affect human behaviour.
◆ It involves what directs this behaviour towards a particular goal.
◆ Motivation concerns how this behaviour is sustained (supported).

For the purpose of reward in connection with maintenance function, care should be exercised to avoid under-awarding or over-awarding. In each case, the system must be analyzed separately to restore equality among the workers.

Restoration of equality: It is possible through the following.

Under-reward: Each one of the following should be examined.

1. Ask for raise
2. Lower input
3. Rationalize why you get less than others
4. Change your comparison with other workers

Over-reward: Here also the same questions must be answered.

1. Try to get raise for other workers
2. Raise input
3. Rationalize why you get more than others
4. Change your comparison with other workers.

IMPLICATIONS OF MOTIVATION THEORIES

It is desired that the maintenance managers should create a congenial work environment that is responsive to individual need, which can give ultimate satisfaction to the workers. However, the following points must be taken care of:

Performance: It will be reflected as per the needs of any employee in the form of job satisfaction or blocked need; activated need.

Understand individual difference: It is important to understand the basic difference between the working personnel. It is well known fact that all jobs cannot be done by all the workers.

Interaction: Manager should be aware of the needs of their workers and know the same that can be taken care of in due course of time.

Knowledge management: It is most important for any body to have full information for successful completion of work and efforts should be directed to maintain harmony during the work environment.

Performance and productivity: This will decide whether motivational efforts are required or not because overmotivation may cause serious consequences.

Employee recognition program: Many a time, it is noticed that if the efforts of any worker are not properly recognized, he feels isolated and finally gets de-motivated towards his job. Therefore, a simple pat from the boss at times will keep the momentum going with fruitful results.

Motivation in Practice

The following steps can be taken in order to motivate the workers practically.

1. *Employee involvement programme:* The employee must feel that he is a part of the system where he is working which is possible through—

   (i) *Participative management:* A process in which subordinates share a significant degree of decision-making power with their immediate superiors. This is required in case of complex job situations.

   (ii) *Representative participation:* It is a process where workers participate in organizational decision making through a small group of representative employees. The basic objective here is to redistribute power within the organization, putting the worker at par with the management in the form of stakeholders.

2. *Work councils:* It include a group of nominated or elected employees who must be involved in decision-making.

3. *Board representatives:* This is also a form of representative participation where, workers are appointed as a board of directors in the interest of the their employees.

4. *Quality circle:* Here a group of employees work together with the management on the quality of products. The group makes concrete suggestions for the improvement of the process or methodology wherever needed.

5. *Employee stock ownership plan:* In this case, the employees are sharing the benefits accrued from the sale of the product or commodity.

6. *Goal setting:* Here work is assigned to the employees based on their qualification and experience in advance for given time period.

7. *Variable pay programme:* Part of the payment is based on the performance of the individual. In this case earning of the employee fluctuates with the measure of performance as against the fixed salary. This may include piece-rate plans, profit sharing plans, gain sharing etc.

8. *Skill-based pay plans:* It is based on the skill of the employee and keeps the motivation in a person alive, to learn more and more for personal gains. This system provides ample chances to the employees to opt out of different benefit programs depending upon their own needs.

SPECIAL ISSUES IN MOTIVATION

It is evident that the process of motivation will be different with the class of people, i.e. professional, workers semi- or skilled, or workers involved in repetitive type of work.

Professional: The basic difference between the professionals and non-professionals is their education pattern. Professionals are more committed to their field of expertise and are more loyal to the profession as compared to the employer. Therefore, it is important for them to keep the information of latest affairs of their fields possibly through Internet, workshops, training seminar/conferences etc. The professionals are more focused on their work as compared to the non-professionals. They may also need autonomy of work, to keep their interests activated. The reward of their work should be in the form of recognition of the work rather than financial gains.

Part-time workers: The workers who are engaged in temporary jobs would like to be made permanent as a part of motivation to them. It is normally assumed that the temporary workers will work hard with the hope of becoming a permanent employee. When temporary employees work along with permanent employee who earn more with extra benefits are bound to be demotivated and finally work has to suffer. This aspect must be taken care of to eliminate discrimination among the workforce.

Skilled workers: The main problem faced by the industry is to how to motivate the persons who are making very low wages and have limited scope to improve it. Such people are normally less educated but have achieved skills through experience. In such cases higher payments only can motivate the persons.

Repetitive job-workers: Such jobs are boring and stressful and therefore care should be exercised for their motivation. One way of motivating persons of this category is to provide clean and attractive work surroundings, number of breaks and socialization with other colleagues during the breaks. The other method could be the compensation of the job through extra payments.

With above all in practice, motivation can be brought in, through timely and proper training of the personnel in their respective work areas including maintenance.

MAINTENANCE TRAINING

In the past, the operator could handle the maintenance work himself as the equipment used was of simple design. With the advancement of technology, automation, electronic controls, and sophisticated mechanisms have been introduced. To maintain such equipment it is necessary to train and develop personnel for this type of work.

Today's maintenance engineer needs a totally new set of skills and considerable knowledge in a variety of complex fields such as electronics, instrumentation, control systems, computers, data acquisition systems, non-destructive testing, robotics, etc. Hence they require better education to be able to understand the constant upgradation of technology and to handle the equipment, its functioning, controls and its preventive maintenance requirements, including overhauls, troubleshooting analysis and calibration.

Training is a continuous process and needs to be adequately planned. All the organizations normally establish their own training departments to impart training at all levels. This programme is usually included under the long-range planning of the company, and however, the short-term training programmes are also be arranged to update the working skills of the maintenance personnel to meet the challenges faced by the industry. In the cases, where training departments are non-existent, the training can be arranged from outside agencies.

TRAINING PROBLEMS

Besides assessing the needs of maintenance training, some other questions also need to be answered. These include the following:

    (a) Does the quality of other maintenance work appear adequate to production supervisors?

    (b) Are equipment breakdowns promptly diagnosed and remedied or help from the outside agency is required?

    (c) Is there a great dependence on certain individuals for specific job knowledge?

If the above questions are not answered satisfactorily, then there is a need for maintenance training which usually covers two areas:

    (a) General and specific craft knowledge

    (b) Special knowledge of a particular plant

General Knowledge

For this purpose, training programmes are arranged at various places in training schools or the organization itself. Although the persons dealing with machines have technical knowledge, they may lack knowledge in maintenance aspects. Therefore, *manpower audit* is to be carried out in an organization to pinpoint the need of maintenance training. The general practice had been to provide on-the-job training to the personnel, besides classroom lectures. In some organizations, it is observed that senior executives undergo training and thereafter-proper transfer of training to the workers is not done. It may also so happen that by the time the equipment arrives at the shop, the trained manager has been either posted elsewhere or promoted to the higher position. It is therefore imperative to impart training to a small group of personnel who are actually responsible for maintenance of the equipment.

Special Knowledge

The process of acquiring special knowledge of a particular plant or equipment is more acute in industry, where rapid technological changes take place. For acquiring special knowledge about an equipment, the maintenance manuals, schematic diagrams, wiring diagrams, hydraulic circuits, etc. need to be available for the maintenance work. For example, if a particular circuit for an equipment is not available, then it becomes very difficult to carry out any repair work effectively. The persons involved in the work have to resort to the practice of hit-and-trial which may not be successful all the times.

TRAINING PROGRAMMES

Depending upon the requirement of skills at different levels, the following training programmes are widely used in industry.

Apprentice Training

To achieve the long-range improvements in basic maintenance skills and to provide continuing help to organizations, apprentice training programmes are mostly conducted by the government organizations. These programmes combine classroom teaching with the on-the-job training conducted over a period of time. Such programmes include shop interactions, blueprints reading or drafting, and study of various maintenance aspects of the equipment. Here training is provided in a general sense as far as the use of tools, machines, or instruments is concerned.

Short-range Programmes

Where specific needs occur in certain special skills, short-range programmes are organized to update the knowledge of maintenance personnel. This is done in the form of classroom studies combined with the on-the-job training. For the specialized works persons must be trained at the company level.

Concurrent Training Programmes

In addition to the long-range type of training programmes such as the apprentice-training programme, many concurrent training programmes like the specific training courses to correct the identified deficiencies, specific training courses to enhance the technical capabilities can also be arranged within or outside the organization. These days, some special programmes are being conducted by the agencies that are dealing with the products.

Passive Training

The training to the workforce can also be imparted passively by properly displaying appropriate figures, instructions at strategic places. Such displays not only educate the workers regarding the right way of doing a particular job, but also motivate them to carry out a job correctly. The captions displayed in work area or outside, do motivate the persons, for example; school ahead, drive slowly, just to keep the accident rate down.

TRAINING FACILITIES

After the training needs have been identified, the available training facilities such as training schools, public or government, vendor schools, study programmes, programmed instruction and in-plant training need to be evaluated in terms of the requirements identified. In our country, mostly in-plant training is adopted, as the other types of facilities are not always available. Nowadays, some facilities are also available for training personnel in some commercial type of equipment and processes, such as computers and NC machines.

INCENTIVES IN MAINTENANCE

It is observed that the majority of personnel involved in maintenance work perform much below the required levels. This is because of management's failure to motivate the persons or to provide the requisite training to accomplish the day-to-day work. Some organizations provide financial incentives to secure high productivity in maintenance work, thereby reducing the total maintenance cost. Overtime payments can be just one type of incentive. Some rewards for better maintenance work should also be instituted by the organizations. An incentive plan, based on measured standard times, requires the development and installation of a cost control function, and planning and scheduling related to maintenance.

Incentive Principles

The incentive plans have to be based on practical data, otherwise they may turn out to be failures. Wherever incentives have achieved success, the two basic reasons have been the properly designed plans and effective administration.

An effective incentive plan cares for better utilization of available manpower and facilities. It should promote good labour relations and reduce personal problems and encourage initiative. It should also be fair to all concerned and remain flexible to accept any changes required in future. A successful plan should give full recognition to the needs of the employees, company, and the public at large. The incentive plans must not contradict the motto of the company, which is to achieve high production without compromising the quality and safety of the personnel and equipment.

Incentive Plans

The maintenance incentive plans would depend upon the type of maintenance work. These plans can be divided into two groups—direct plans based on the measurement of maintenance work, and indirect plans based on other factors such as production output and revenues earned.

Direct plans

These plans may be based on one or more of the types of measurements such as job estimating, work sampling, past performance and standard data with quality factors to fix norms for incentive payments and are applicable to direct maintenance work carried out by the personnel engaged in the shop. This work may be of routine type such as overhaul of components/parts, etc.

Indirect plans

These plans are based on maintenance cost including the cost of labour, materials, and the overhead expenses. Several indices are used in combination as a basis of incentive payment. During a specified period, if the availability of the equipment increases, leading to increase in production, and the cost of maintenance work is also brought down, the concerned persons can be suitably rewarded for boosting the morale of the workers. This type of plan can also be applied to persons who are working as service personnel.

LABOUR PERFORMANCE ANALYSIS

It is easier for the maintenance engineer to control the material and overhead costs than the labour costs because the nature of work is unknown in most of the cases. Therefore, it is important that cost analysis be carried out in respect of the maintenance labour force. This can be done in terms of the manhour utilization or the work output in the form of reports available to the management to improve labour performance. The maintenance cost as compared to the production cost can be another measurement.

Control reports show the results of actual performance and relate these actual conditions to standard performance. These reports also permit fast action to improve the maintenance performance by eliminating the causes responsible for poor performance. Sometimes the nature of work allocated may not suit a particular worker, dissuading him not to take interest in his work. This eventually results in poor performance; under such situations persons with forward lookout must be employed for better results.

Control reports also give the details such as weekly labour analysis, weekly performance summary, individual employees performance reports, labour performance trends, and maintenance performance trends and thus highlight the indirect factors responsible for poor maintenance work so that remedial actions can be suggested in time.

MEASURING PERFORMANCE IMPROVEMENT

In the case of maintenance work, it is difficult to measure performance as compared to other parameters in production. However, some basis has to be fixed up for the purpose of comparison and for evaluation of the consequent improvements in performance. The labour utilization can give an indication of labour performance. The indices used for this purpose include:

- ◆ The number of men in the maintenance group
- ◆ Maintenance labour payrolls
- ◆ The measured labour performance

Of the three indices measured, labour performance is the best indicator of any change in maintenance performance. Performance improvement is indicated as the percentage improvement of the current performance. For example, in a shop, the number of components/parts repaired/serviced in a particular period will indicate the improvement.

Improving Day-to-Day Performance

Day-to-day performance of labour has to be evaluated individually or group-wise and the same can be improved through the following ways:

1. Publishing the results of performance of individuals or a group, which will stimulate the workers to improve their own performance.
2. Competition between groups will also help in improving labour performance, provided it is fair and done carefully.
3. The names of the individual/group who have topped in performance can be displayed to encourage the workers.

Rewards in the form of prizes for better performance will help in further improving and encouraging labour performance. However, shortfalls if any should also be projected to avoid reoccurrence of the event.

Improving Performance through Incentives

While measured day-work control will increase the productivity, the use of incentive plans will further increase it. Incentives used in the form of extra pay, for extra effort/work is an effective way of improving labour performance. The results from incentive programmes can be maximized by demolishing all factors, which interfere with the output of the employees towards better performance.

Maintenance-Based Compensation

Since jobs of maintenance personnel are of different nature, due compensation should be given to them at par with other personnel of the production department. The concept of contract can also be applied in maintenance work if it is of non-repetitive in nature. This helps the management to reduce the downtime of the equipment. If the work is completed before the stipulated time period, incentives can be given, whereas for delays, penalties can be imposed. For this reason only, many industrial units prefer contract type of work in the areas such as computers, air-conditioners, photocopiers, etc.

Each person is an individual, whether a mechanic or chief engineer of the plant. For fulfilling the objectives of the organization, it is necessary to account for every individual's role in the department for effective utilization and accounting. No one should be treated as a burden on the other.

Economic Aspects of Incentive Plans

In an organization where the number of employees engaged in maintenance work is more, the management may be reluctant to accept any incentive plan. For any plan, therefore, maintenance cost per unit time must be calculated without and with incentive plans to highlight the benefits from incentive plans. The cost of the administrative work required to implement the incentive programme includes functions which should be done without any incentive plan for the effective working of the maintenance organization. Efforts should therefore be made to reduce the cost of administrative work by using computers to handle a large quantity of data for setting standards of job times. In some areas computer-based systems the maintenance costs can be reduced appreciably.

The amount incurred towards any incentive programme is an important parameter for the total maintenance cost and should be given due attention. It should be based on the amount of extra work done and be in the form of percentage of the salary. Though the practice of production bonus is followed in all industrial organizations, the same has not proved to be that successful towards motivating the workers. There have to be real incentive programmes either to enhance the maintenance work or to motivate the workers. While adopting the incentive plans, the following facts must be considered:

1. Some administrative measures should be adopted for control of costs, whether following an incentive plan or not.
2. An efficient administrative system should be made the part of the maintenance function before deciding upon any incentive programme. The incentive plan must take care of work and skill levels of the individuals involved.
3. The supervisor and workers should together show cohesiveness in their working. Every individual involved should know in advance about the standards set in the organization.
4. If there are unions in the organization, they should be taken into confidence while fixing up the incentive programmes.
5. The mode of payment for incentives should be carefully selected keeping the working conditions in mind, i.e. nature of job.
6. The use of ratio factors for the purpose of incentive plans should be avoided and the incentive should be directly related to the output, skill and the efforts of the workers and the requirements of the maintenance work.
7. Whole-hearted support of the management and the employees is essential for the successful application and adoption of any incentive programme.
8. Persons involved in the implementation of the incentive programme should cooperate among themselves for efficient working.
9. The competent and knowledgeable persons must be employed for conducting the time study and rate setting of the jobs.
10. As far as possible, the piece rate of incentive plan should be adopted and not the hours of working because some jobs take more time as compared to others.

QUESTIONS

1. How does motivation affect the performance of individuals in an organization? Explain.
2. Explain the motivation process briefly.

3. List down the characteristics of motivation.

4. Write down the basic motivational theories used in practice.

5. What are the implications of motivation? Explain.

6. What are the special issues in motivation? Discuss briefly.

7. Enumerate the training problems that are faced during performance of the maintenance functions.

8. What are the different types of training programmes that can be imparted to the maintenance workforce?

9. Discuss how incentives can improve the performance of a maintenance organization.

10. Labour performance analysis can enhance the productivity of maintenance personnel. Justify.

11. Examine the economic aspects of incentive plans used in the maintenance.

7

Computers in Maintenance

INTRODUCTION

For the optimum performance of a plant, effective resource management and reliable equipment are the essential factors. The Computerized Maintenance Management System (CMMS) is employed to keep track of these factors. This system not only helps in defining changes in infrastructure, management philosophy and employee skills but also assists in implementation and selection of the programme.

For the effective discharge of the maintenance function, a well-designed information system is an essential tool. Such systems serve as effective decision-support tools in maintenance planning and execution.

Today, there is an increasing trend of application of information-based decision-support systems in different departments of modern industry. Thus, computers have become an indispensable requirement in maintenance management. Maintenance planning is needed for the preparation of maintenance schedules and job distribution. The need of manpower planning, at this stage cannot be overlooked. The planning must take care of organizational goals and policies and then formulate procedures for effective implementation.

For optimum maintenance scheduling, a large volume of data pertaining to men, money and equipment is normally required to be handled. This obviously is a difficult task to be performed manually. For a planned and an advanced maintenance system, adequate and timely information is a must, which is possible only by the use of computers.

The computer is an efficient and reliable tool for maintenance personnel to plan and implement their programmes. However, the success of a computer-based system solely depends on the way the system is planned and the way the data are processed. The inputs to the computer system are exactly the same as the inputs to a man-made system. Programmes can be prepared to have the available inputs processed with the help of computers. Such a computer-based system can be used as and when required for the effective performance of the maintenance tasks.

The following are the main advantages of using computers in decision making for maintenance functions:

- ◆ Large quantity of data can be stored
- ◆ Response time is less
- ◆ Better accuracy of information
- ◆ Operation and research methods can be used effectively
- ◆ Cost optimization
- ◆ Reduction in paper work
- ◆ Minimization of information leakage
- ◆ Easy feedback of information
- ◆ Number of alternative solutions are possible
- ◆ Accurate forecast and better planning

COMPUTER-AIDED MAINTENANCE

Temporary stoppages and shutdowns of some of the essential industrial systems such as the power plants, fertilizer plants, petrochemical plants, etc., can be very disastrous; attention must, therefore, be focussed on systematic maintenance planning. This planning is quite complex in nature and needs an objective approach to decision-making, execution and control. For these, it is essential to have accurate and up-to-date information in an organized manner. The industries mentioned above have a very large number of complex systems; thus it is very difficult to monitor their maintenance requirements. Without a proper computer-based system, the decisions made may be sub-optimal and erroneous. There are a wide variety of software packages available in the market for different types of maintenance systems.

Computers enable implementation of online monitoring of important equipment. This online concept is widely used in condition monitoring, where the equipment performance is checked at a fixed interval, and any deviation in the performance is investigated for verification of the fault. In the case of a flexible manufacturing system or an automated system, some maintenance functions such as change of a broken tool is automatically done with the help of a program provided in the computer system. With the advancements taking place in the field of artificial intelligence, the risky maintenance jobs such as those encountered in steel plants and the likes can be easily programmed to be performed by robots.

A computerized maintenance system includes the following aspects.

- ◆ Development of a database to support utilization of periodic maintenance programmes
- ◆ Analysis of past records if available, to ensure the frequency of maintenance programmes
- ◆ Development of maintenance schedules
- ◆ Availability of maintenance materials and manpower
- ◆ Feedback control system for the assessment of the programmes carried out
- ◆ Project management to achieve the set objectives of the maintenance department along with the organizational goals

In a large-sized industrial organization, the following computer-based maintenance systems can be implemented.

Development of Database and Analysis

In any maintenance system information about failure frequency of the parts/components is essential. The data collected must be authentic and accurate as far as possible. Their suitability, reliability, adequacy etc. must be ascertained, since all other decisions will be based on it. During analysis if the failure distribution is not clear, non-parametric tests must be done to test the hypothesis. In case where information is not available, the same can be generated, following a particular distribution as applicable to the situation. However, proper testing of the generated data must be carried out to establish the failure/repair time distributions to match the real life conditions. Otherwise the results obtained will not be accurate and it will be different from the real life situations.

Development of Maintenance Schedule

With the available information, it will be possible to generate maintenance schedules to know the maintenance requirements such as manpower, material etc. As the maintenance function by itself is most unpredictable, care should be taken to keep time and manpower for untimely breakdowns of the machines. This can effectively be dealt with once dynamic scheduling is followed. Due care should also be taken to avoid backlogging of the equipment during any maintenance policy being practiced.

Job Card System

It is essential to prepare a job card for each equipment/component to record the maintenance work carried out or the work to be done. The job card shows the plant code, the equipment code, the job code, the nature of the jobs, the time of initialization of the card, the start time of the job, the finish time of the job, the man-hours spent, and so forth. These filled cards are sent to the computer section for processing and necessary compilation. The plant engineer uses the feedback provided by the computer for compiling the following information:

- History of maintenance
- Frequency of maintenance operation
- Average maintenance time for various jobs
- Capital required equipment-wise for the maintenance function
- Manpower required category-wise for the maintenance function.

The use of computers facilitates the issue of job cards, recording of job history, and control of manpower, etc. These filled cards are sent to the computer section for processing and necessary compilation.

Maintenance Material Planning

This is one of the very important factors of the maintenance function. With the help of a computer-aided maintenance system, it is possible to know the precise material requirements in advance and therefore the procurement process of quality products can be initiated well in time. The details of the spare parts required can be stored in the computer either equipment-wise or item-wise. With the available forecasting techniques the exact quantity can be known in advance and therefore, procurement becomes easy. The information from central store to the site stores can also be made available now with the facility of Internet. A material master file is created for all equipment/systems under a particular organization.

Feedback System

The preparation of job cards helps the maintenance personnel to know the exact position of each maintenance-related job in time, in the form of reports received from the computer system. These reports can be matched with the actual jobs to find out any deficiency. The feedback control system helps the maintenance engineer to redesign or reschedule the maintenance activities to make proper use of the facilities and material. This will also help in the following functionalities.

- Providing accurate number of spare parts
- Reduction in non-moving lot of spare parts
- Maintaining the optimum level of spare parts inventory

Use of Network Methods

For all equipment, detailed networks showing the time for each maintenance activity can be drawn up in consultation with the maintenance personnel. This information is then circulated among the plant engineers to analyze the shortcomings in the activities, if any. The list of activities, which can be delayed, is also prepared separately. With the help of the network, the completion time of each maintenance job can be easily calculated to help planners to make the maintenance schedules.

Project Management

The above listed activities are to be managed by a group for efficient working with some conflicting situations in terms of the cost of maintenance versus availability of the equipment/facility for human cause. This must be resolved amicably without burdening either the user or the organization concerned. For the purpose of management, the use of computers can help to a greater extent, where prompt decisions can be taken, based on information available. Each maintenance work should be treated as a mini project.

Maintenance Cost System

To know the overall maintenance costs, the month-wise details of cost of maintenance in respect of each plant can be prepared indicating labour cost, material cost, and overhead expenditure. The actual cost is then compared with the budgeted cost. This also helps in deciding the replacement policy of the equipment based on the maintenance cost being incurred for the old equipment. Operating cost of the equipment is also reduced/minimized with proper maintenance policy.

Computer Applications in Inventory Control

The main purpose of inventory control is the monitoring of stock levels and is achieved by recording stock movements on stock records. Through this inventory position can be known in advance including shortages, if any. The maintenance work generally suffers due to non-availability of spare parts in time. Forecasting spare parts and other relevant materials can be improved by the use of computers. The details of the spare parts can be stored either equipment-wise or item-wise. If this work is done manually, it consumes a lot of time and also the information may not be so accurate as desired. Therefore use of computers is inevitable. A material master file is created to keep the record of each item, equipment-wise and this record is updated whenever some change takes place during the maintenance work. The inventory control is necessary for the following reasons:

- ◆ Avoidance of excessive stock of spare parts
- ◆ Order placement cost optimization
- ◆ Optimization of inventory carrying costs
- ◆ Minimization of lead-time

Spare Parts Life Monitoring System

Under this system, information about a spare part such as its description, anticipated life, and the date of its installation in an equipment, and so forth, is usually recorded. As and when a particular spare part is replaced during breakdown failure or scheduled maintenance, the updating of this information is done in the respective files stored in the computer. With this available information, it helps to prepare the following reports.

- ◆ Spares parts repeatability, in various machines, indicating the performance of each spare parts with the details of its failures, if any.
- ◆ Comparisons of the actual life, with the estimated life of the spare parts as indicated by the manufacturers.

Spare Parts Tracking System

The total time required to rectify a breakdown of a machine is the sum total of the times required for each of the following operations:

- ◆ Time to identify the cause of the failure
- ◆ Time to determine the requirement of spare parts
- ◆ Time to procure the spare parts
- ◆ Time to rectify the failure
- ◆ Testing and certification of equipment/machine

In most of the cases the maximum time is consumed in procurement of the spare parts when the same is not available in the store. With the application of the computerized system, the use of spare part tracking system is beneficial in this respect to a great extent. A spare part file is created that contains the information about the material code; spare part identification number, the assembly or sub-assembly number and the place where the spare part is used. This helps in knowing the current position about a particular spare part and facilitates a timely procurement for future demands. A desired level of spare parts inventory can be effectively maintained.

Material Requirement Planning (MRP)

We have seen that forecasting the requirement of spare parts can be improved by the use of computers. All the relevant details about the use of a spare part can be maintained in the computer for ready reference and placement of the requisite order in time. The use of past consumption records of spare parts helps in planning future maintenance functions. The other facilities required can also be planned in advance since their time of use is known. This also helps to avail the services of the experts in the area of maintenance for complex and specialized equipment/services where the downtime of the systems is of importance.

Predictive Maintenance versus Computers

Normally the manufacturer provides a checklist to carry out the routine inspection of the equipment. The preventive maintenance schedules can be drawn using computers and the checklist and follow-up actions can be done through the feedback system.

If the information is fed into the computer properly, priorities can be set in respect of the pending jobs. This also helps in proper manpower planning for additional work, if any. The status of each activity can be monitored as and when required. The proper scheduling and rescheduling of the jobs can also be done.

The aim of such predictive maintenance, therefore, is to identify the critical parameters of an equipment or plant that can be continuously monitored for the corrective action. Hence, predictive maintenance helps in pointing out potential failures well in advance. When the number of critical parameters is more or when the rate of monitoring the critical parameters is high, the use of computers is most desirable.

The critical parameters such as vibration, temperature, and so forth, which need to be continuously monitored, can be recorded on the online monitoring system interfaced with the computer. Through this data, the base line signature of a particular system/component can be maintained in the computer file. The signals received from the working system can be compared with the baseline signature. Any deviation in the measured value of the parameter beyond the tolerance limit is highlighted for necessary remedial action. This total procedure is called *signature analysis*. Similarly, monitoring the equipment condition for corrective actions through a close-loop control system with the help of computers is also possible.

The implementation of the predictive maintenance depends on the type of the system and its importance in the organization. Any standby system will call for such a monitoring. The computer-based systems can be implemented for the following purposes:

♦ Monitoring critical parameters through a data acquisition system
♦ Monitoring critical parameters through a feedback system

- Establishing a breakdown analysis system
- Computer simulation in maintenance

When a number of similar equipment is used in any industry, the computerized breakdown analysis helps in procuring the inventory and planning the preventive maintenance schedules. This system aims at generating important details for the purpose of breakdown analysis and updates the equipment history as well.

Equipment log sheets are used for collecting information about the equipment performance, breakdown durations, reasons for breakdowns, and corrective actions taken. This information is updated at regular intervals of times. A master file is maintained to know the requirements of spare part.

The following useful reports, for example, can be obtained by the use of a computerized maintenance system:

- A complete list of breakdowns with reasons, for all equipment, on a weekly/monthly/yearly basis
- Repetitive kinds of breakdowns, indicating the frequency of failures
- Meantime between failures and the expected repair time, equipment-wise
- Spare parts repeatability in various machines with their performance in each machine
- New spare parts, life monitoring

A number of software packages are available in the market to deal with the inventory management details, which are available through Internet and published materials. Most of the micro-based software are interactive and menu driven.

Maintenance Information System

The function of maintenance information system is to convert field data into useful information for effective decision-making on maintenance issues. A key element in the improvement of equipment reliability is the quality information, presented in a logical and understandable format. This system should also provide relevant information for day-to-day maintenance work. The information required include labour control, backlog, maintenance cost besides repair history of the equipment with defined performance indices.

Information is power and should be utilized properly for any system and particularly for the maintenance system since the success of the whole system depends upon the accuracy and a timely availability of information. Such information can be used to establish the objectives and to direct the attainment of these objectives by comparing the actual results with the objectives. This information also helps in decision making, if collected properly. Therefore, it is important that the collection of information be done accurately and in the right manner.

The monitoring of labour control information is vital to ensure that the labour resources are efficiently allocated and effectively utilized. It also enables the maintenance department to measure and control over time usage. Another measure of proper workforce size and composition is the maintenance backlog. An excessive backlog will mean that the modifications to the size, composition and/or distribution of the workforce must be made.

The cost information retrieved from the system provides a measure of the ability of the maintenance department to perform the required work within the specified budgetary provisions. This information should be available at various levels including fleet, equipment and component levels. At the fleet and equipment levels, the information provided will help to decide the maintenance policy of overhaul or replacement of a particular equipment to make it more cost effective. At the component level, the high costs would indicate that the repair history should be reviewed and possibly that corrective maintenance is required. The maintenance or repair history provides vital information on repairs made to key production equipment over a selected time period.

This must be analyzed to identify patterns of chronic, repeated failures, which require corrective maintenance action, and component failure causes and life cycles to establish PM change out schedules and to develop the most effective equipment service programme. The repair history must be monitored on an on-going basis to ensure that any corrective maintenance actions and modified practices are effective in prolonging component life.

The type of information to be collected depends upon the type into: (a) decision-making information needed to identify problems, develop solutions, and corrective decision to be made. In the case of maintenance functions, the information required can be broadly categorized actions, control the actions and measure performance, and (b) administrative information which includes route data reports, open work order listings or scheduling of routine and preventive mainte-nance inspections. Performance indices such as maintenance cost per tonne are reviewed regularly and coordinated directly with elements of decision-making information. The information system must also be capable of generating reports on defined maintenance performance indices. Some example include:

- ◆ Maintenance man-hours or cost per tonne produced.

  Percentage of maintenance man-hours or costs attributable to each of emergency and unscheduled repairs, PM services and inspections or planned and scheduled maintenance work; and

◆ Percentage compliance with weekly maintenance schedule.

These indicators, and others, will be vital for monitoring the progress of the maintenance departments in achieving a more proactive maintenance focus.

Maintenance information can be used for issue and monitoring of job orders and work instructions. Numerous vendors provide such packages, which must be customized for a particular site of application.

The information in respect of maintenance functions can be collected at various levels as indicated below.

Unit level

After understanding the need and importance of maintenance, the required items of information are listed out which must form a part of maintenance department's database for computerization. As the repair work must be done promptly and efficiently, the information about breakdowns must be collected at the unit level. Such a database helps in procurement of adequate quantities of spare parts particularly of breakdown-prone parts. The frequency of a breakdown must be known for predicting preventive maintenance schedules. The information collected at the unit level must include costs such as, idle time cost of equipment, material cost, labour cost, and so forth. This information helps to provide optimal maintenance.

Aggregate level

Proper compilation of information at this level will help the maintenance management system to get reorganized. Information collected is presented in the form of monthly/quarterly/annual reports. These reports are analyzed on a common basis. Sometimes there is no uniform understanding of terminology, for example, the downtime of equipment. In some units, if the maintenance work is done on holidays or Sundays or even in the third shift, this time is not compiled towards the equipment downtime and should be accounted for during collection of information.

COMPUTERIZED MAINTENANCE MODELS

The role of computers is significant where condition based maintenance system is practised. At the same time importance of accurate information is vital for maintenance function, as well as, to optimize the overall performance of the equipment/system. With the help of vibration analyzer the condition of some of the parts of mechanical equipment can be monitored at a fixed time interval. This will help to draw maintenance schedules accurately and timely. Some of the modern equipment is provided with warning switches, which can indicate impending failures and are also used as safety measures. In the case of on-line monitoring system, critical parameters such as temperature, pressure are recorded regularly which is interfaced with computer.

Though the manufacturers recommend a maintenance schedule for their products, but random failures do take place. Information about such failures can be collected and analyzed to help in building up a more authentic maintenance schedule. Where such real data is not available, computer simulation can be used to generate artificial data of the same failure pattern that is likely to occur in a system. For this purpose, breakdown and repair time data of each critical part/component of equipment need to be carefully analyzed for establishing the following:

(a) *Breakdown distribution:* This will indicate the time between failures of the equipment.

(b) *Breakdown period:* A distribution of the length of time during which the equipment remained out of service.

(c) A distribution of the *length of time* for the actual repair time of the equipment.

MAINTENANCE DECISION MAKING

Maintenance management clarifies the difference between preventive maintenance and breakdown maintenance. The preventive maintenance can be planned but it is difficult to plan the breakdown maintenance since its occurrence is unknown. Through experience it is established that the proper planning and scheduling of preventive maintenance can save time and money and therefore overall maintenance cost reduction. It is a well-established fact that the use of computers can immensely save time and reduce the cost of maintaining records. The Transactions Processing System (TPS) have looked at the records of components, spare parts, fixtures and tools as inputs and consumption statements and maintenance accounts as outputs. Though these inputs are still valid for building a corporate management information system, the following additional data can also be included:

- ◆ Equipment conditions
- ◆ History of failures
- ◆ Direct cost of maintenance
- ◆ Inventory values and material movements
- ◆ Man-hours spent on maintenance
- ◆ Overtime paid, usage of other facilities made
- ◆ Performance of the maintenance workforce
- ◆ Reliability and maintainability of equipment, and maintenance cost per unit of production achieved.

From such database, maintenance reports can be generated for equipment control such as equipment register, equipment history, major failure reports, forecast of maintenance, plant non-availability, plant reliability, maintainability, and maintenance schedules. The reports like work, control, craft performance, maintenance performance, overtime, delay in work can also be generated from this database. The idea of cost control, material transaction, and the likes can also be made available to the maintenance department through such a database as and when required.

Making decision is a critical element of any organizational life and most vital in the case of maintenance function. It provides very less time to decide the course of action to be contemplated. In case of major breakdown, it is essential to restore the equipment/system to the working condition as quickly as possible. One way could be to replace the whole system for the time being if the same is available and carry out other operations in usual time. The situation of the breakdown of equipment is very important.

The frequency of failures can guide the maintenance managers to adopt the suitable policy to restore the equipment in working order. Decision regarding replacement versus maintenance will also depend on the cost involved and the facilities available for the repair of the failed equipment. Non-availability of spare parts and sometimes the use of old equipment itself forces the management to use the equipment after repair/overhaul.

In general, some decision-making models are available in the literature, but their direct use in the case of maintenance functions is difficult due to time constraints. However, the same can be used where policy decisions for maintenance planning etc. are to be decided. Creativity plays an important role in decision making and therefore, efforts should be made to improve it since it provides novel and useful ideas.

There may be differences in decision making by individuals because, the people differ along two dimensions. The first is their way of thinking, where people are logical and rational. They process information serially. On the other hand, some people are intuitive and creative. They perceive things as a whole. The other dimension addresses a person's tolerance for ambiguity. Some people have a high need to structure information in way that minimize ambiguity, while others are able to process many thoughts at the same time. When these two dimensions are grouped, four distinct decision making styles emerge which are directive, analytical, conceptual and behavioural.

Persons using directive style have low tolerance for ambiguity and seek rationality. On the other hand, analytical style people have greater tolerance for ambiguity and need more information for decision-making.

Individuals with conceptual style tend to be very broad in their outlook and consider many alternatives whereas, in the case of behavioural style people would like to work together. This model can be effectively used in case of maintenance functions since the work is not individual based; rather it is associated with group activities.

Expert Systems

The expert systems are computer programs built for commercial applications using the programming techniques of artificial intelligence, especially those techniques developed for problem-solving. There are a number of expert systems available for using in the maintenance operations. Building of expert systems is concerned with knowledge engineering, which is a sub-discipline of artificial intelligence. Expert systems have been built for medical diagnostics, electronic fault finding, mineral prospecting, condition monitoring, and so forth.

The above discussions reveal the basic need of computers in the maintenance department. As the technology is advancing fast, sophisticated equipment are being produced everyday, which calls for improved reliability in maintenance. Since reliability is important in every field, its importance in the area of maintenance cannot be overlooked, and therefore needs proper attention.

COMPUTERIZED MAINTENANCE PLANNING

As planning is one of the important functions of maintenance and requires considerable time, the use of computers can be made to ease the job. The maintenance planning is basically done to optimize the available resources. The works like, issue of work order, priority of work, material planning, work plan, job sequence etc. can be handled effectively with accuracy. It is therefore a powerful tool for assisting maintenance planning. A well conceived computerized maintenance planning should provide manpower backlogs, equipment histories, equipment part list, material availability, preventive maintenance schedules and associated costs.

Effective maintenance planning requires observing planned job progress to eliminate potential delays or problems that may arise. During the follow up process the planner may address the problem areas, which are:

- ◆ Were the communications clear and adequate for the people associated with the work?
- ◆ Was the job plan simple enough to understand?

The basic reason for planning maintenance is to eliminate the cause while overcoming the effects efficiently. The benefits of computerized maintenance planning include the following:

- Better access to the information
- Better planning of available resources
- Better control and
- Overall cost reduction

The initial high cost of the computerized maintenance planning can be compensated in the due course of time and the usage of the system. The strategic issues associated with computer-aided maintenance while decision-making are the following:

- *Communication gap:* There is a tendency to create communication gap between the computer professionals and the user-manager, which restricts its use.
- *Computer-vendors role:* It is often noticed that the software engineers try to influence the users by indicating that their product has solutions for all type of problems.
- *Master plan:* It is seen that the effective master plan minimizes the computer failures, which are due to hardware acquisition, software development and MIS design etc.
- *Management information system:* Looking into multipurpose use of computers it is required that the same may be provided to all the executives with Internet facility for the better interaction.
- *Managerial participation:* After ensuring management support it is desired that the user participation should start in the design phase on corporate MIS so as to avoid subsequent extensive and time consuming re-work.
- *Information needs:* In most of the cases, it is observed that the concerned managers fail to identify the basic needs of the information in the specified area.
- *Human acceptance:* It is noticed that people are averse to change though it may yield many fruitful results and introduction of new system requires lots of persuasion.

Computer in Human Resource Management

The significance of computers in the area of human resource management is also essential where the number of manpower utilized is high and the same is incorporated in fulfilling the objectives of the organization. The information requirements within the broad framework include the following areas:

- To support wage and salary administration
- Recruitment and appointments
- For organizational promotion policy
- For training programme of employees
- Potential development and identification of working force

The human resource management information systems are required for three main purposes:

1. To store personal details of individual employees for reference
2. To provide basis for effective decision-making
3. To provide information to other departments/organizations

A comprehensive system of records covers all the information required about the employees but it should be authentic, adequate and relevant. It is necessary to avoid any gaps in information, which is essential for decision-making at the same time storing/collection of useless information be avoided. Regular reviews of the stored records must be made to ensure that they will serve the useful purpose with minimum cost.

QUESTIONS

1. Explain how computers can be helpful in discharge of maintenance functions.
2. How can inventory control be acquired with the use of computers?
3. List the computer-based maintenance systems which can be practised in maintenance functions.
4. Briefly describe the maintenance information system used for maintenance functions.
5. At what levels can the information be collected for maintenance functions.
6. How can simulation be applied to the maintenance function? Explain briefly.
7. Briefly enumerate the information needed for making decisions in maintenance functions.
8. How can expert systems be applied to maintenance function? Explain briefly.
9. Describe computerized maintenance models used in practice.
10. Explain the role of computers in human resource management.

Chapter

8

Reliability in Maintenance

INTRODUCTION

A complex industrial system uses the sophisticated equipment. The use of these equipments has increased the demand to identify the failure-prone parts and to define planned maintenance programmes. The concept of reliability is not very old and had started with the U.S. Army Department in the 1950s, where a group was formed to study the reliability of the critical parts and to predict quantitatively the performance of the equipment. This was done with the idea to improve the military missions who were failures due to poor operational conditions of the equipment used.

The growing need for reliability arises from the requirement to continuously run the equipment in a system in the best possible manner. Higher reliability of equipment would, therefore, ensure uninterrupted availability of the system to the users. Thus, reliability of equipment is of prime importance to maintenance personnel, since they have to keep the equipment in a state of readiness at all times for carrying out its designated functions effectively.

Introduction of reliability into a system is, therefore, basically a design and manufacturing feature that precludes or minimizes failure by implementation of strict quality control measures, meant to increase the life and operational availability of equipment. The main advantages of imposing reliability requirements are, therefore, increased productivity, reductions in forced outages of equipment because of planned maintenance work. The application of reliability techniques to any equipment helps to identify the flaws in system design, compare several possible system configurations, minimize downtime, maximize operational readiness, reduce operating costs, and develop maintenance policies. The maintainability aspects of the systems must also be cared during the design stage.

The assessment of equipment reliability during operation and deployment gives an idea of replacement of its components as stipulated at different time intervals. It also helps in planning the scheduled maintenance work to enhance the equipment availability for production/service work. If the overall reliability of equipment is assessed low, it suggests that emphasis should be placed on periodic maintenance work; otherwise the components prone to wear fail at a faster rate will further bring down the equipment reliability. Thus the prediction of maintenance needs, to ensure desirable level of equipment reliability for achieving the production targets, is of great importance to maintenance personnel.

RELIABILITY

Definition of Reliability

Reliability is defined as the probability that a component/system, when operating under given conditions, will perform its intended functions adequately for a specified period of time without interruptions. It refers to the likelihood that an equipment will not fail during its operation. Reliability can be of two types:

- ◆ Inherent reliability
- ◆ Achievable reliability

Inherent reliability is associated with the quality of the material and the design of the machine parts along with its processing methodologies. On the other hand, achievable reliability depends upon other factors, such as maintenance and operation of the equipment. In order to increase the reliability of a system, provision of redundancy within the system or use of standby equipment can take care of contingencies in the event of failures. This implies the use of more than one item or equipment for a given system and, therefore, needs cost-benefit analysis before implementation of the decision of increasing the reliability of the system by this method. It is also possible through better maintenance practices.

The quantitative assessment of reliability of a given system is a cumbersome process and it requires the following steps. First the system is divided into various subsystems and a block diagram is then constructed. The reliability is evaluated through development of a reliability models as discussed in the following sections.

Reliability and Probability

These two terms are closely associated with each other, as reliability means the probability that a system will operate for a given period without a failure.

Let

R be the probability of reliable function for a specified period of time

F be the probability of failure occurring during the same period of time

Then, obviously

$$R + F = 1$$

Failure Rate (λ)

The number of failures occurring in a unit time is known as the failure rate, λ. It can be calculated during a specified period for a given number of components. If there are 5 failures occurring during a period of 1000 hours to which 500 components are subjected, then

$$\lambda = \frac{5}{500} \times \frac{1}{1000}$$

$$= 1 \times 10^{-5} \text{ failures/hour}$$

Failure Rate Estimation

Failure rate estimation is carried out by using the available failure data of the components/units concerned.

The failure rate calculation of the floating pinion and link unit (FPLU) from the field outage data is presented in Table 8.1.

Table 8.1 Failure rate of FPLU

C.I. (days)	Mid value, x	Frequency, f	fx
0–4	2	18	36
4–8	6	10	60
8–12	10	13	130
12–16	14	9	126
16–20	18	7	126
20–24	22	5	110
24–28	26	4	104
28–32	30	4	120
32–36	34	2	68
36–40	38	2	76
40–44	42	2	84
44–48	46	1	46
48–52	50	1	50
52–56	54	0	00
56–60	58	1	58
60–64	62	1	62
		Σf 80	Σfx 1256

From Table 8.1, the mean time to failure is given as

$$m = \frac{\Sigma fx}{\Sigma f} = \frac{1256}{80} = 15.7 \, \text{days}$$

$$= 376.8 \, \text{hours}$$

Thus the failure rate of the FPLU is

$$\frac{1}{m} = \frac{1}{376.8}$$

$$= 2.7 \times 10^{-3} \, \text{failures/hour}$$

Failure Pattern of Equipment

The failure pattern of an equipment over its whole life cycle can be represented as shown in Figure 8.1. Here, Phase-I shows the failure pattern inherent in a new product because of manufacturing or design defects. This period is also called the *infant mortality period of equipment*. Here, the number of failures is high, but they are not of serious nature.

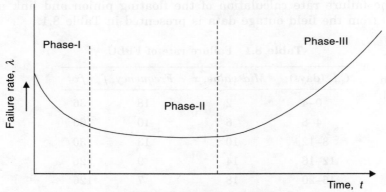

Figure 8.1 Equipment life cycle.

Phase-II shows the useful life period of an equipment where the failure rates are normally moderate as the equipment gets set to the working environment. The last phase of the equipment life is Phase-III, and the failures occurring during this phase fall under the category of wear-out failures that are caused due to aging of the equipment.

The above parameters of the equipment life cycle are the essential requirements for the prediction of system reliability. The other important parameters such as repair time distribution can be used to estimate availability, maintainability and level of corrective and preventive maintenance. Before developing any reliability model it is essential to carry out failure data analysis. This can be done from the available failure data of the components/equipment. It will help to develop a reliability model effectively. The various models used for reliability assessment are as follows:

Series reliability model

In this model, the components are arranged in series and the success of the system depends on the success of all its components. Consider the reliability model shown in Figure 8.2 in which n components having reliabilities as R_1, R_2, ..., R_n are connected in series. The reliability of the complete system would be

$$R_S = R_1 \times R_2 \times R_3 \times \cdots \times R_n$$

$$= \prod_{i=1}^{n} R_i$$

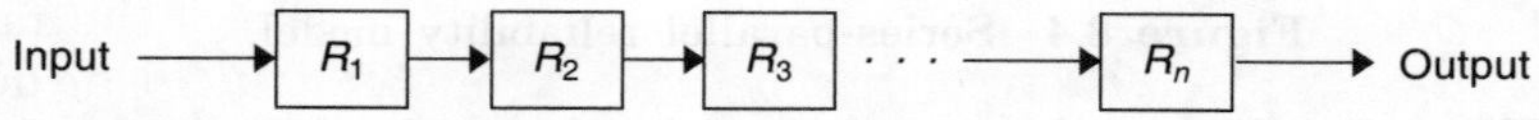

Input ⟶ R_1 ⟶ R_2 ⟶ R_3 $\cdots$ ⟶ R_n ⟶ Output

Figure 8.2 Series reliability model.

In this configuration, the failure of any component puts the complete system in down position.

Parallel reliability model

In this model, the system can be partially operative even if some of its components are in the failed state. Let R_1, R_2, ..., R_n be the reliabilities of n different units used in the system as shown in Figure 8.3. Then the product law of unreliability can calculate the total system reliability as given below:

$$R_S = 1 - \prod_{i=1}^{n} (1 - R_i)$$

$$= 1 - \prod_{i=1}^{n} \left(1 - e^{-\lambda it}\right)$$

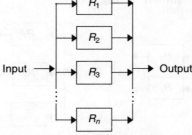

Figure 8.3 Parallel reliability model.

Series–parallel reliability model

In practice, more complex configurations exist where components are arranged in series and parallel. The reliability structure of such a complex system is decomposed into simpler structures through successive application of the conditional probability theorem. The technique starts with the selection of a key component, which appears to bind together the overall probability structure, and the reliability may then be expressed in terms of the key components as shown in Figure 8.4.

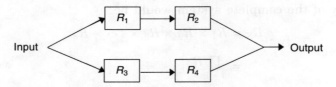

Figure 8.4 Series-parallel reliability model.

The reliability of the system can be determined from the failure probability of the above arrangement as a consequence of $R + F = 1$, or $R = 1 - F$. Considering the above arrangement, let R_1, R_2, R_3, and R_4 be the reliabilities of units 1, 2, 3, and 4 respectively. Failure of R_1, R_2 is expressed as (1 – probability of reliability).

Thus,

$$\text{Failure of } R_1, R_2 = 1 - R_1 R_2$$
$$\text{Failure of } R_3, R_4 = 1 - R_3 R_4$$

and

$$R_S = 1 - (1 - R_1 R_2)(1 - R_3 R_4)$$

The above relation can be simplified as

$$R_S = e^{-(\lambda_1 + \lambda_2)} + e^{-(\lambda_3 + \lambda_4)} - e^{-(\lambda_1 + \lambda_2 + \lambda_3 + \lambda_4)}$$

Redundant reliability model

In a system consisting of some identical components connected in parallel to provide higher reliability, the calculation of system reliability is done as shown in Figure 8.5.

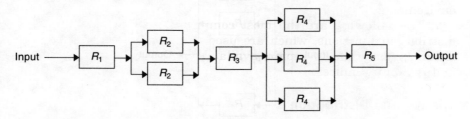

Figure 8.5 Redundant reliability model.

Let R_A be equivalent to:

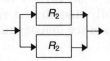

Let R_B be equivalent to:

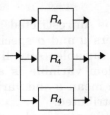

Thus, substituting the equivalent components, we have:

$$R_A = 1 - (1 - R_2)^2$$

$$R_B = 1 - (1 - R_4)^3$$

Hence the reliability of the whole system is

$$R_S = R_1 \times R_A \times R_3 \times R_B \times R_5$$

and the equivalent series system will be as shown in Figure 8.6.

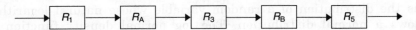

Figure 8.6 Equivalent series system.

FAILURE FUNCTIONS AND THEIR MODELS

Because of automation of industrial systems, it has become essential to identify the failure-indicating parameters before the actual failure of the equipment/part takes place. These methods are termed hazard/failure functions and are of great help in reliability analysis and in determining the failure density functions and failure distributions. Failures can occur due to various reasons such as material wear-out, poor maintenance, bad designs and unexpected loads, and working environment.

The following are the most commonly used failure and repair distribution functions, which are used to calculate the mean time to failure, mean time to repair, availability, maintainability functions, and reliability indices.

The Normal Distribution

It is the most important continuous probability distribution and useful in countless applications. It describes quite accurately the random variables associated with a wide variety of experiments.

The range of a normally distributed random variable consists of all real numbers. The probability density function (p.d.f.) is defined by the equation

$$f(x) = \frac{1}{\sigma\sqrt{2\pi}}\, e^{-(x-\mu)^2/2\sigma^2} \quad \text{for} \ -\infty \leq x \leq \infty$$

where the parameter μ is unrestricted and the parameter σ is positive. The two parameters μ and σ specify the mean and standard deviation of the random variable. Any linear transformation of a normally distributed random variable is also normally distributed. That is, if x is normal with mean μ and variance σ, and if $y = a \cdot x + b$, then y is also normally distributed.

The density function of a standard normal random variable is

$$f(x) = \frac{1}{\sqrt{2\pi}}\, e^{-x^2/2} \quad \text{for} \ -\infty \leq x \leq \infty$$

The normal distribution is reproductive, that is, the sum of two or more normally distributed random variables is itself normally distributed.

The Log Normal Distribution

It is the distribution of a random variable whose natural logarithm follows a normal distribution. The log normal density function is expressed as

$$f(x) = \frac{1}{\sigma x\sqrt{2\pi}}\, e^{-(\ln x - \mu)^2/2\sigma^2}$$

The range of the random variable is all $x > 0$.

The parameters μ and σ indicate:

$$\mu = E(\ln x)$$
$$\sigma^2 = v(\ln x)$$

But the mean and variance of x are expressed as

$$E(x) = e^{\mu + (1 + 2\sigma^2)}$$
$$Y(x) = e^{2\mu} + \sigma^2\left(e^{\sigma^2} - 1\right)$$

The log normal distribution arises from the product of many independent non-negative random variables. This is in contrast to the normal, which arises from the sum of independent random variables. The log normal distribution has been used to describe the lifetimes of mechanical and electrical systems in engineering applications.

The Gamma Distribution

Let x be a continuous random variable defined over the range $x \geq 0$. It expresses gamma distribution if the density function is of the form

$$f(x) = \frac{\lambda^r x^{r-1}}{\Gamma(r)} e^{-\lambda x} \geq 0$$

where λ and r are positive and $\Gamma(r)$ is the gamma function defined by

$$\Gamma(r) = \int_0^\infty x^{r-1} e^{-x} \, dx$$

When r is an integer, the gamma distribution reduces to the Erlang.

The Weibull Distribution

A continuous random variable defined over a range $x \geq 0$ has a Weibull distribution if the density function has the following form:

$$f(x) = \lambda \beta (\lambda x)^{\beta-1} e^{-[\lambda x]^\beta} \geq 0$$

where λ and β are positive constants. When $\beta = 1$, the density function reduces to negative exponential distribution. The mean of a Weibull distributed random variable is

$$E(x) = \frac{1}{\lambda} \Gamma\left(1 + \frac{1}{\beta}\right)$$

and the variance is

$$V(x) = \frac{1}{\lambda^2} \left\{ \Gamma\left(1 + \frac{2}{\beta}\right) - \left[\Gamma\left(1 + \frac{1}{\beta}\right)\right]^2 \right\}$$

This distribution has its application in describing lifetimes and waiting times in reliability functions. If a system consists of a large number of parts each of which has a lifetime distribution of its own (independent of others), and if the system fails as soon as one of the parts does, then the lifetime of the system is the minimum of the lifetimes of its parts. Under these conditions, there is theoretical justification for expecting a Weibull distribution to provide a close approximation to the lifetime distribution of the system.

The Poisson Distribution

Let x be a discrete variable defined over the range, $x = 0, 1, 2, 3, \ldots, \infty$

If $p(x) = \dfrac{\lambda^x e^{-\lambda}}{x!}$ with $x = 0, 1, 2$

then x has a Poisson distribution with parameter λ with positive values. It has many properties and therefore finds a wide use in modelling. The expectation and variance are equal to one another. That is,

$$E(x) = V(x) = \lambda$$

This distribution is reproductive, which means that the sum of Poisson distribution random variables will be another Poisson distributed random variable.

One of the common uses of Poisson distribution is an approximation to the binomial distribution when the number of trials is large, while the probability of occurrence is small.

The Negative Exponential Distribution

If x is a continuous random variable defined over the range $x \geq 0$, then

$$f(x) = \lambda e^{-\lambda x} \qquad \text{for} \quad x \geq 0$$

where λ is a positive parameter, x has negative exponential distribution or sometimes positive exponential distribution. The expectation of negative exponentially distributed random variable is

$$E(x) = \frac{1}{\lambda}$$

and variance is the square of the same value, that is,

$$V(x) = \frac{1}{\lambda^2}$$

This distribution is widely used to describe random variables that correspond to duration. In other words, it is the waiting time distribution process. The above information will be helpful in determining the frequency of failures, mean time between the failures and mean time to repair a failed unit using the relevant distribution.

RELIABILITY APPLICATION

To improve the availability of an equipment/system, failure analysis must be done to predict the equipment/system reliability. Most of the variable characteristics of a practical system are random in form or can be represented by an appropriate random distribution. The reliability model of a system is essentially a model derived from the concepts of probability and statistics under certain assumed conditions. The actual model depends upon a large number of factors that must be known to the analyst during the formulation process. Some of these factors are as follows:

♦ System configuration/functional arrangement
♦ Types of variables which may influence the performance aspects of the various elements of the system such as failure/success
♦ Model formulation.

For the reliability analysis of any system, the system must be divided into functional elements composed of units/subsystems/components, and a block diagram of the system operation is constructed too. When the element probabilities are independent, calculations are easy but difficulties are encountered when the number of elements involved is large and several approximations for combination rules of reliability need to be applied. The model becomes complex when some of the elements are dependent and in such cases, joint density and distribution functions or Markov Chain may be needed to solve the problems.

The following example of a crusher used in the coal processing plants shows how the calculations of reliability are made (Figure 8.7).

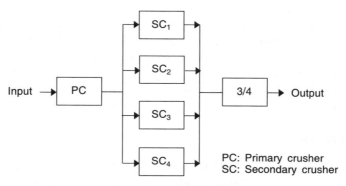

Figure 8.7 The crushing equipment.

For reducing the ore size, different types of crushers are used. The crushing equipment comprises a primary crushing plant (coarse crusher called the rotary breaker having 600 tonnes/hour capacity, 152.4 mm feed size and 127 mm set size) in series with a secondary crushing plant. The plant requires three out of its four secondary double roll crushers (of 200 tonnes/hour capacity each) to be running to produce the desired fineness of the product (37 mm).

The primary crusher is highly reliable; it is the secondary crusher that is prone to frequent failures and has been the subject of intensive study in the past years.

The crusher reliability problem needs to be taken into account when constructing an adequate model from the field data on failure and/or repair of its key components. Failure profiles of the key components, which contribute to the successful performance and operation of the crushing equipment, are studied. This study enables to predict the quantitative reliability of the crusher. From the reliability of each individual crusher, the system reliability of the crusher equipment is calculated. The comparison of system reliabilities obtained at different times of usage makes it possible to replace the components at the right interval of time, plan for preventive maintenance schedules, and maximize the equipment operational availability in order to increase the production capacity.

The secondary crusher system consists of four gearmatic double roll crushers. Each crusher has the following major subsystems:

◆ Motor and drive unit (MDU)
◆ Flywheel and shaft unit (FSU)
◆ Driver pinion and gear unit (DPGU)
◆ Floating pinion and link unit (FPLU)
◆ Moveable roll (MR)
◆ Fixed roll (FR)

The motor and drive unit (MDU) consists of an induction motor fitted with a pulley which drives the flywheel through a V-belt. The flywheel and shaft unit (FSU) transfers the flywheel power through the jack shaft to the driver pinion and gear unit (DPGU). The latter at one end drives the fixed roll (FR), and at the other end through the floating pinion and link unit (FPLU) it drives the moveable roll (MR). The distance between the rolls is adjustable and the two-roll assembly is supported on a heavy and rigid cast fabricated frame.

Reliability Model

The most important consideration in reliability evaluation is the development of a reliability model based on the failure and/or repair behaviour of the functional units of the system under study. Usually the functional units are put together in a block diagram to constitute a series and/or parallel model.

The basic reliability equation for a single unit operating in the useful life period where the failure rate is constant, is given by

$$R(t) = e^{-\lambda t}$$

where

λ is the failure rate

$R(t)$ is the probability that the unit will continuously perform its intended function in the time interval $0-t$.

In case the units are connected in series, the product of the individual unit reliabilities gives the reliability of the system. Thus,

$$R_S = \prod_{i=1}^{n} R_i = e^{-\sum_{i=1}^{n} \lambda_i t}$$

The unreliability of the series system is

$$F = 1 - R$$

The case of the parallel redundant units is dealt within a different manner because the system remains partially operative even if one of the units is working well. In this case the reliability of the system is evaluated as follows in terms of the products of the individual unit unreliability:

$$F = \prod_{i=1}^{n} F_i$$

and

$$R = 1 - F$$

Therefore,

$$R = 1 - \prod_{i=1}^{n} \left(1 - e^{-\lambda_i t}\right)$$

The case of at least r out of n identical units can be handled using the following relationship.

$$R\left(\frac{r}{n}\right) = \sum_{k=r}^{n} {}^nC_k e^{-k\lambda t}\left(1 - e^{-\lambda t}\right)^{n-k}$$

The general assumptions in the reliability evaluation model include that:

1. The units are independent.
2. The failure of one unit does not influence the other units.
3. The probability of occurrence of failure of two or more units simultaneously is negligible.
4. The time during which a unit is replaced by an identical unit is neglected.
5. Each unit has an exponential failure distribution curve.
6. The failure rate is constant during the useful life of a unit.
7. The reliability of a unit at $t = 0$ is unity and it thereafter decays exponentially with time.
8. The probability of a unit failing in the useful life period is attributed to chance failure.
9. The maintenance outage of a unit is considered as the unit failure.

The reliability model for the gearmatic double roll crusher is shown in Figure 8.8.

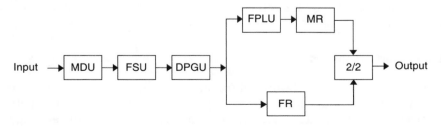

Figure 8.8 Secondary crusher reliability model.

Reliability Evaluation

The reliability model of secondary crusher is shown in Figure 8.8. It is redrawn using its major subsystems as described earlier. It is then reduced to that shown in Figure 8.9 by combining the various functional units.

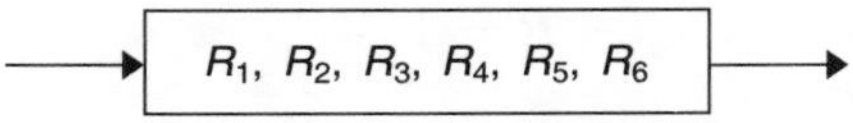

Figure 8.9 Modified secondary crusher reliability model.

Thus the reliability $R_C(t)$ is evaluated as

where $R_C(t) = R_1 \times R_2 \times R_3 \times R_4 \times R_5 R_6$

$\quad R_1 = e^{-\lambda_1 t}$ = reliability of the MDU

$\quad R_2 = e^{-\lambda_2 t}$ = reliability of the FSU

$\quad R_3 = e^{-\lambda_3 t}$ = reliability of the DPGU

$\quad R_4 = e^{-\lambda_4 t}$ = reliability of the FPLU

$R_5 R_6 = e^{-(\lambda_5 + \lambda_6)t}$ = reliability of the MR – FR combination

In recent years, the quantitative reliability techniques are being increasingly used to predict the performance of equipment. The procedure involves establishing the failure behaviour of the components in the given system. The success of the reliability evaluation procedure is largely dependent on the outage or failure data collected from the field. Table 8.2 shows the failure data collected over a period of five years. Using a simulation technique, the failure rates of the various functional units can be calculated as demonstrated earlier in this chapter taking the case of the FPLU.

The secondary crusher system reliability in terms of the individual crusher's reliability is evaluated as

$$R_{CS}(t) = R_C^3(t)\left[1 - R_C(t)\right]4 + R_C^4(t)$$

Table 8.3 gives the reliability values of different subsystems of the crusher.

Case Study Analysis

Table 8.2 shows the failure data of the major subsystems of the crusher. Using this data, the subsystem reliabilities as a function of time were calculated. The crusher as well as the crusher system reliabilities as obtained at every interval of 2 hours of operation in a day (generally having two shifts of 8 hours duration each) are presented in Table 8.3.

Table 8.2 Failure data for a secondary crusher subsystem

Unit	Symbol	Failures/hour
MDU	R_1	0.30×10^{-3}
FSU	R_2	2.20×10^{-3}
DPGU	R_3	2.40×10^{-3}
FPLU	R_4	2.70×10^{-3}
MR	R_5	0.72×10^{-3}
FR	R_6	0.40×10^{-3}

Table 8.3 The reliability values of different subsystems of the secondary crusher

T(h)	R_1	R_2	R_3	R_4	R_5	R_6	$R_C(t)$	$R_{CS}(t)$
0	1.0000	1.0000	1.0000	1.0000	1.0000	1.0000	1.0000	1.0000
2	0.9999	0.9956	0.9952	0.9946	0.9986	0.9992	0.983	0.998
4	0.9998	0.9912	0.9904	0.9893	0.9971	0.9984	0.957	0.990
6	0.9982	0.9869	0.9857	0.9839	0.9957	0.9976	0.949	0.985
8	0.9976	0.9826	0.9810	0.9786	0.9943	0.9968	0.933	0.975
10	0.9970	0.9782	0.9762	0.9734	0.9928	0.9960	0.916	0.962
12	0.9964	0.9739	0.9716	0.9681	0.9914	0.9952	0.901	0.952
14	0.9958	0.9676	0.9670	0.9629	0.9900	0.9944	0.885	0.932
16	0.9952	0.9654	0.9623	0.9577	0.9885	0.9936	0.870	0.915

The analysis of the above information reveals that:

(a) The reliabilities of all the subsystems such as the motor and drive unit (MDU), flywheel and shaft unit (FSU), driver and pinion and gear unit (DPGU), floating pinion and link unit (FPLU), moveable roll (MR), and fixed roll (FR) are different at different times of usage.

(b) The overall reliability of the secondary crusher system, $R_{CS}(t)$, is greater than the reliability of the individual crushers, $R_C(t)$, at any operational hour.

(c) $R_{CS}(t)$ is greater than $R_C(t)$ as the crusher system has the 3/4 redundancy of the crushers. The system capacity is 600 tonnes/hour, whereas each individual crusher has a capacity of 200 tonnes/hour. This means that one of the crushers serves as a standby unit.

(d) The probability of outage of two or more units of crushing equipment at a given time is negligible because of effective maintenance efforts of a preventive type at scheduled intervals of time. From Table 8.3, it is obvious that the system reliability drops down to 0.915 in 16 hours of continuous use as against 0.975 in 8 hours. This, in fact, is a warning situation and emphasizes the importance of schedu-led maintenance during the two production shifts (of 8 hours duration each), apart from one maintenance shift (of 8 hours duration) in a day.

(e) The shift-wise preventive maintenance, if carried out care-fully and effectively, can help to find out the deterioration taking place in a unit before it actually fails and can thus minimize the breakdown outage duration.

The reliability of the crushing system as used in the coal processing plants is a good illustration of the measure of its successful operation during the production schedule.

The most important consideration in reliability evaluation is a complete awareness of the modes of failures and operation of the system under study. Once the system is understood, it is possible to divide the system into various subsystems, and subsystems into various functional elements.

From the functional units, a reliability model is constructed and using the outage data of various units the quantitative performance of the overall system, referred to as the reliability, is predicted.

It is evident from Table 8.3 that the comparatively large drooping reliability of RFSU, RDPGU, and RFPLU with respect to operational hours calls for frequent maintenance care during produc-tion shifts everyday.

After 16 hours of continuous use, the reliability of the crusher system falls down to below 90 per cent, therefore, it becomes essential that the unit undergoes a thorough maintenance check before it is put into service again. If maintenance is not carried out carefully after the continuous running of two shifts, the units would be prone to failures at a faster rate, resulting in a sharp fall in reliability.

However, the aforementioned case study analysis is of immense use in predicting the maintenance needs in order to ensure a desirable level of the system reliability satisfying the production objectives. The method explained here can compare the crushing equipment system design alternatives as well.

The above discussions show the importance of reliability and the maintenance functions, which go hand in hand. For improving the reliability of a system/equipment, proper care must be taken to the maintenance of components. For this purpose, the relevant facilities, including special tools and spare parts should be made available to the maintenance department in time.

DESIGN FOR RELIABILITY

Owing to the requirements of higher reliability it is desired that the concept of reliability must be introduced at the level of design. The main requirement of design for reliability is that the independent checks are carried out to ensure that the 'paper' design meets the specifications. A team of experts can study the affects of all variations under all working conditions and failure rates of the components. This process concludes with the design review board, which makes the final decision to approve the design. The design test carried out will provide the following information on:

(a) The basic individual causes of failures, and the condition responsible for the same.
(b) Number of failures due to each basic cause as a percentage of the total number of overall failures.

The general principles of design for reliability include the following:

- ◆ *Component/element selection:* The components/elements used should confirm to the specifications and their failures must be within the limits. The established technologies only should be used to maintain the reliability standards.

- ◆ *Factor of safety:* It is known that the premature failures can occur if the parts/components are subjected to over loading. This must be taken care of during the design phase with the selection of proper factor of safety. For mechanical parts a safety margin of better than 5 should be used and in case of electronic circuit the voltage stress rate be kept below 0.7 to minimize their failures.

- ◆ *Environment:* It is observed in practice, that the frequency of failures also depends on the working environment and therefore these factors must be considered, specially for critical components/parts to minimize failures.

- ◆ *Complexity of system:* It is seen that in case of series reliability model the reliability depends upon the number of components in a system. Whereas, in case of parallel reliability model it is path dependent of the components. Attempts should be made to have proper combination of components so that the reliability of the equipment/system is highest. In case of electronic systems use of integrated circuits be made to minimize the failure rates.

- ◆ *Redundancy:* In case of series reliability model the failure of a component makes the system inoperative and therefore it is desired, that as far as possible parallel model must be developed particularly for critical items.

- ◆ *Diversity:* If the probability of common mode failure limits the reliability of the overall system, the use of equipment diversity should be considered. Here, a given function is carried out by the systems in parallel but each system is made up of different elements with different operating principles. The example is temperature using electronic and pneumatic systems, which may have different failure pattern.

◆ *Reliability calculations:* From failure rates of the components/elements the overall system reliability can be calculated depending upon their configuration, i.e. series, parallel or combination of the two. The calculated reliability/ failure rate for the overall product should then be compared with the target value, if this value is not found within the specified limits, the design should be modified till the target figure is reached.

Reliability Indices

The use of indices for reliability is very common to assess the performance of the equipment/system. Some of the indices are based on the probabilistic considerations. However, the new generation of indices provide a fruitful comparison of the performance based on the collected statistics and also predict the future system performance. The following indices are used:

$$SAIFI = \frac{\text{Total number of customer interruptions}}{\text{Total number of customers served}}$$

$$SAIDI = \frac{\text{Sum of customer interruptions durations}}{\text{Total number of customers served}}$$

$$CAIFI = \frac{\text{Total number of customer interruptions}}{\text{Total number of customers interrupted}}$$

$$CAIDI = \frac{\text{Sum of customer interruption durations}}{\text{Total number of customers interrupted}}$$

$$ASAI = \frac{\text{Customer hours of available service}}{\text{Customer hours demanded}}$$

where;

SAIFI: System Average Interruption Frequency Index
SAIDI: System Average Interruption Duration Index
CAIFI: Customer Average Interruption Frequency Index
CAIDI: Customer Average Interruption Duration Index
ASAI: Average Service Availability Index

QUALITY AND RELIABILITY

Quality of a product can be defined as its ability to ensure complete satisfaction to the customer/user for which he has paid. This is also associated with reliability, which means that a quality product will have higher reliability but not all the times because of variability in production process. For maintaining quality some techniques are used to keep them under the specified limits. It is observed from the failure analysis that high quality products have less number of failures as compared to poor quality. The achievement of high quality not only improves the reliability of the equipment but also adds to its value. During the calculation of reliability, the reliability of each component is taken into account and therefore, the impact of quality is vital and significant. The quality products also draw attention of the users/ customers, as the maintenance problems are less.

RELIABILITY IMPROVEMENT

As discussed earlier the requirements of reliability are more demanding in case of costly and risk attached equipment/systems. When new equipment is to be purchased the procurement procedure, specification of item, vendor selection and bid conditioning etc. play important role. The other methods for reliability improvement include load conditions, attitude of material supplier, spare parts and their impact on the service life of equipment, modern maintenance techniques. In some cases, special seals for pumps and motors can improve reliability and safety of the system. As it is seen in practice that no equipment can work trouble free, irrespective of any maintenance system but the reliability of the machine can be improved by effective and timely maintenance. In general the concept of standby unit is practiced to maintain the high level of reliability for any system. The various approaches used to improve the reliability in the design phase are as follows:

- ◆ The user's needs be examined critically to minimize the use of unrealiable parts/components
- ◆ Use of standby unit to minimize interruptions
- ◆ Selection of quality parts
- ◆ Avoidance of overloading of machine
- ◆ Control of working environment
- ◆ Proper monitoring of maintenance practices
- ◆ Use of robust design methods
- ◆ Screening test to identify infant age failures
- ◆ Research and development for unreliable parts and components.

RELIABILITY TESTING

Why reliability testing: Testing of any manufactured/fabricated product is must, but with the concept of better reliability, its importance grows manyfold specifically for items where risk factor is high. The testing requirements for the products/systems depend on the place of usage, environmental conditions, expected failure mode, reliability standards set and the expected life. The other parameters considered for reliability testing include sample size, which include: single sampling; double sampling and multiple sequential sampling can be used depending upon the quantity and quality of the product. For maintaining uniformity certain predefined reliability standards are set during the design phase. The testing should approximate to the actual field environment and must always be repeatable.

To meet out the design and reliability requirements it is essential to test the product to minimize the failures. The problem of testing is significant in case of live items particularly electronic and electrical. The accelerated tests are therefore carried out for the primary reliability-related factors of applied voltage, temperature and humidity. In addition, statistical sampling is used, taking into account the similarities between products in process and design, so as to minimize the number of test samples. Reliability testing should be considered as a part of integrated test programme which include the following:

1. *Statistical testing:* It is done to optimize the design of the product and the manufacturing processes.
2. *Functional testing:* It is conducted to confirm that the design meets the performance requirements.
3. *Environmental testing:* It is done to ensure that the designed product can work successfully under the given conditions.
4. *Reliability testing:* It is performed to ensure that product renders services for the specified period of time without undesirable problems.
5. *Safety testing:* It is done to ensure all safety aspects has been taken care off.

Reliability is the function of time and variability therefore; the tests should be carried out for a long period to ensure failure distributions, particularly wear out failures. The important consideration in reliability testing is that test results must contribute to the evaluation and improvement of product reliability.

Reliability Tests

These tests are carried out to compare the fabricated/manufactured products to meet the design requirements as per specifications. The target failure rate within the limits under all conditions must be verified. A full reliability test may comprise the three basic activities.

(1) **Failure rate:** Under specified conditions of temperature range, stress and environmental conditions the failures of the items must be measured on individual component or on random samples. This may involve testing at different level of temperature, humidity, mechanical vibration, mechanical shock operating voltage and salinity.

(2) **Burn-in:** The above-cited conditions of testing are made more severe than the defined specifications. This is done to accelerate the failure of weak or defective components/parts so that they fail during tests. The main aim here is to minimize the early failure region as shown in bathtub curve, so that the useful life of the product is extended.

(3) **Screening:** This includes visual inspection, constant acceleration and measurement of electrical parameters of components and assemblies.

Reliability Data Sources

The data required for reliability analysis can be obtained from various sources, which include the following

◆ Data from laboratory where simulated conditions are applied.
◆ Field data on similar parts or components obtained during controlled development or operational test programme.
◆ Field data from similar parts or components obtained during normal operational use.
◆ Field data from generically similar parts or components used in variety of applications.

The data thus obtained may not be accurate for the following reasons:

◆ Inadequacies of the data forms in an effort to compromise between administrative expediency and technical comprehensiveness;
◆ Erroneous or incomplete data due to the complexity of causes and mechanisms, time restraints in reporting, inadequacy of personnel training;
◆ Ambiguity of replies, omission of statements of fact committing persons responsible for primary causes of failure arising from improper operation or omitted maintenance;
◆ Incomplete descriptions, particularly of components, failure definition and description;
◆ Inability to provide accurate operating times.

In the developed nations like the USA, Japan, Germany and USSR Systems Reliability Services help the industrial organizations in this regard and provide accurate and authentic information about failure of parts/components used.

The confidence, which can be assigned to any failure data is dependent upon the nature and source of the data and the number of samples from which it has been obtained.

QUESTIONS

1. Explain how the reliability data helps in performance of the maintenance function.

2. What is reliability? Discuss various reliability models used in practice for maintenance of equipment.

3. Why does the failure pattern of equipment not remain uniform throughout its life? Explain.

4. Discuss the various distribution functions used for the estimation of reliability in the performance of the maintenance function.

5. A system consists of four components connected in series with failure rates of 0.75, 1.0, 0.1, and 0.1. Compute the probability of their failure after six months.

6. A standby feed water system consists of n identical pumps and a flow diverter. The diverter enables the discharge from any pump to be connected to the output pipe so that the system is seen with one pump operational and the remainder standby. The overall system is to have a reliability better than 0.97 over a period of six months. Use the data given to find the value of n.

   Constant failure rate for a single pump = 0.45 per year. Reliability of a single diverter operation = 0.900.

7. An electrical power system consists of identical diesel genera-tors, each generator has a constant failure rate of 0.5 per year. Calculate the reliability of the following arrange-ments—three months after the initial commissioning date.

   (a) Only one generator operating.

   (b) One generator operating, one generator standby.

   (c) One generator operating, two generators standby.

8. Briefly discuss how the reliability of the equipment can be improved.

9. Explain the importance of quality in context with the reliability.

10. Discuss the important points, which should be included in the list of reliability tests.

11. Briefly discuss the sources from where reliability data can be obtained.

9

Development of Maintenance Engineering Practices

INTRODUCTION

The life and reliability of equipment and components can be improved by the proper application of tribology and terrotechnology. In most of the developed industrial organizations, tribology groups have been formed to undertake studies to solve wear and friction problems. The objective of such an approach in maintenance is to initiate the most appropriate action plan at the right time. The formation of R & D groups helps to a great extent, where similar types of organizations coordinate the activities jointly. A proper information and document-ation system further helps the working groups for the similar types of problems associated with the maintenance function. A typical maintenance system is shown in Figure 9.1. Such a system is further improved by incorporating the tribological and other advanced technical and managerial practices.

TRIBOLOGY PRACTICES

Tribology is the science and technology of friction, wear and tear aspects. It deals with the study of friction and wear, and their control. Industrial studies have revealed that the cost of replacement of worn-out components may be as high as 70 per cent of the total mainte-nance cost. Tribology has gained importance because of its direct influence in reducing maintenance costs and improving equipment performance. An effective management of tribology requires careful considerations of various aspects such as material science (compo-sition of material and heat treatment), accuracy in design and manufacture, lubrication (type and application) and service conditions (such as temperature, humidity, dust, fumes, and the like).

A tribological study of any system involves the determination of the frequencies and the modes of failures caused by wear. This helps

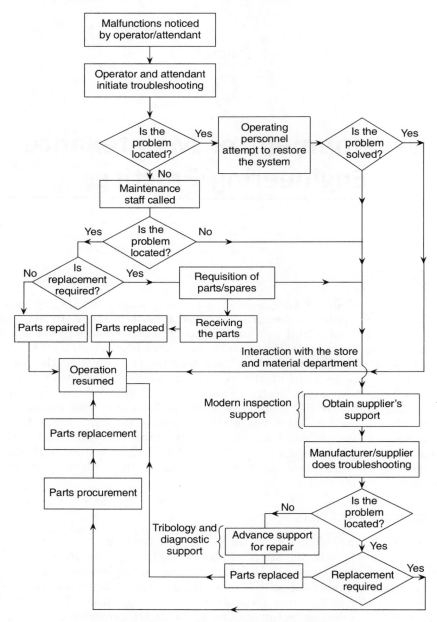

Figure 9.1   Typical maintenance system.

the maintenance engineer to decide if any change in the material of any component is required. This also assists in identifying the type of lubricant that suits the system most. The application of system approach to tribology provides means to identify the problems accurately in a short time, which is otherwise time consuming and tedious. No sooner the problem is known, the remedial actions can be taken without delay.

The abrasive wear is the main cause of component failure where sharp-edged bulk materials are transported in a system. As the material has to be handled protective measures have to be implemented for the protection of the system. Wear is also associated with reduction in operating efficiency, increase in power losses and eventual breakdown of the machine demanding repair/maintenance. An increase in the operating life of a component by overlaying or tero-coating with superior wear resistant alloy increases the reliability of the component. This also reduces downtime and inventories thereby increasing the productivity, and profitability of the organization.

To implement a tribology programme in any industry, working groups, comprising experts from all the concerned departments, are formed. A tribo-analysis report form is designed to collect the detailed information in a quantified manner. The maintenance working groups refer their problems to the tribology group giving full details in the prescribed form. The committee then undertakes the improvement programme considering the importance and cost of equipment. In this way the company evolves solutions to a number of wear-related problems with a view to achieving reduction in overall maintenance costs. Documentation of any tribological study should be done very carefully. Such documents come out to be of great value to other maintenance groups facing similar problems. Some of the results obtained through the application of tribology to maintenance problems are shown in Table 9.1.

Table 9.1 Application of tribology to solve maintenance problems

System	Problem	Solution	Results
Cold shear blade	Chipping	Weld deposit of Co-base electrodes	100 per cent life improvement
Rod mill liners	Wear and breakage	Selection of proper rubber liner	100 per cent life improvement
Impeller fan blade	Fluid erosion	Hardening of blades	No replacement of blades
Wear liners	Abrasive wear	Selection of proper rubber liner	100 per cent life improvement
Hammer mill	Wear and abrasion	Heat treatment	130 per cent life improvement
Bearings	Breakage	Material change	300 per cent improvement
Lubricant	Low utilization	Change in the type of lubricant	100 per cent life improvement
Spherical roller	Wear and breakage	Change of lubricant	200 per cent life improvement
Conveyor belts	Wear on top and bottom covers	Selection of the best type of material	150 per cent life improvement
Rubber components	Wear and abrasion	New designs	Life cost reduction by 300 per cent

TERROTECHNOLOGY PRACTICES

In any industry it is highly desirable that the equipment remains available for the maximum time period for operational utilization. This therefore calls for *designing-out* maintenance philosophy, which is the main theme of terrotechnology practices. The application of this concept requires the following design features to be included in modern equipment.

1. The design of the equipment should be such that it requires no or minimal maintenance work. This ensures better performance and productivity.
2. Provision should be made to send periodic feedback on the performance of the system to the designer to enable him to make modifications to achieve the desired performance.
3. All concerned personnel should be given proper training on working and maintenance aspects of equipment. The normal practice followed is, that the operating and maintenance groups are separated and their coordination is essential.

Though the design-out maintenance is ideal to improve the equipment availability for production work, it is difficult to achieve this standard. Thus the terrotechnology practices provide the means for modifying the existing equipment to improve reliability and also to reduce the maintenance work.

The important areas of terrotechnology are: condition monitoring, protective devices, and application of advance maintenance engineering. Condition monitoring provides assessment of failure of critical parts, which is studied and analyzed while redesigning the equipment for improved performance and higher reliability. Figure 9.2 shows a model that can be used for terrotechnological analysis and improvement.

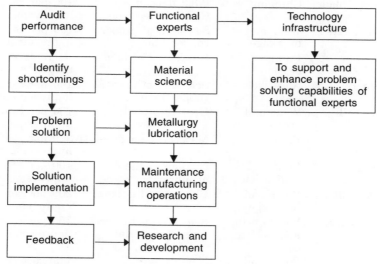

Figure 9.2 Terrotechnology analysis and improvement.

INDIGENOUS SUBSTITUTION

One of the pertinent problems of the maintenance management is to have adequate provision of spare parts. Downtimes of capital intensive machinery due to non-availability of spare parts can jeopardize the economic performance of any industry. In many instances, important machinery suffers from this serious drawback. As a solution to this problem, maintenance management needs to take the adequate steps to develop indigenous spare parts particularly for the critical components, which are not available in the market. To this end, the bigger industries establish and maintain an R & D division in the organization to take care of all the aspects of the product.

Keeping in view the requirement for future development of indigenous spare parts, all the relevant drawings and blueprints must be procured along with the imported machinery. The change of model/design makes the availability of spare parts difficult under above conditions.

In many cases it is seen that imported machinery is procured without the detailed drawings. Later, it causes serious problems in indigenous development of spare parts. For successful development of indigenous items, a collaborative approach proves very effective. Here, all the concerned persons sit together to give their requirements with respect to performance, reliability, and maintainability of the components and assemblies. In our country, defence sector has shown the lead in the development work for indigenous parts and many successful products have been developed to international standards. This approach not only saves foreign exchange, but the development of indigenous technology is also accelerated at the same time. As far as possible, efforts should be made at the organizational level to provide alternatives of the maintenance related problems.

RECONDITIONING

Reconditioning is preceded by technical inspection. The concept of assessing the remaining life is now becoming a problem. Decision-making on reconditioning depends on the expected, required service life of the machine with minimum operating cost for the remaining period.

Reconditioning of equipment at proper times not only improves the performance of the equipment, but also enhances its operating life. During this process, all the subassemblies are thoroughly checked and tested to the standardized levels. The replacement of costly assemblies can thus be avoided. Reconditioning is mostly practised for the maintenance of components where the cost involved is a small fraction of the cost of the new components. Normally during reconditioning, the equipment is dismantled and the subassemblies are sent to different shops for overhauling. If the spare parts required for reconditioning are not made available in time, the maintenance function will be delayed and the downtime of the equipment will also go up appreciably.

ADVANCED STRATEGIES

The role of the maintenance managers has become more important today as they have to ensure higher availability of equipment coupled with high reliability. The requirement of higher availability means reduction/elimination of unplanned shutdowns of the equipment. Therefore, the maintenance manager needs to know the problem in advance so that proper arrangement for the spares can be made in time to avoid delays. All this should be done at minimal cost, keeping in view the quality of maintenance work. Some of the strategies to meet these requirements are discussed below.

Maintenance Information Management System (MIMS)

The need of the proper maintenance information management system arises due to complexity and size of equipment coupled with the requirement to meet the higher availability and reliability targets set by the user departments. The MIMS enables the maintenance engineers to simulate the maintenance-related information in a proper way and at the right time. This information system handles data such as equipment inspection schedules, service and replacement records, spare parts requirements and their procurement, manpower planning, failure analysis, and preventive/scheduled maintenance, and so forth. Depending upon the organizational requirements, this system serves in various ways the different areas of maintenance management such as:

(a) Maintenance schedules generation
(b) Shop control/material movement
(c) Inventory control/spare parts management
(d) Failure history of the component/parts, and
(e) Analysis of maintenance systems for deciding the use of particular scheme.

Maintenance schedules give information about job planning, scheduling and monitoring of all the related works. The function of shop control is related to the work involved in manufacturing of certain parts for maintenance work at the organizational level. This may also include repair of certain items/assemblies, as well as movement of the material and maintenance related equipment and tools.

The most important information needed in maintenance work is the inventory of spare parts. Properly maintained inventory reduces the lead-time of maintenance work. This includes detailed information about equipment and sub-assemblies for procurement of spare parts. It is also necessary at this level that the inventory model to be used must be known in advance for placing orders. In case of manufacture/ fabrication of certain spare part sufficient time must be kept in the programme.

The failure history of the equipment compiled during its use, helps the maintenance personnel to plan the preventive/scheduled maintenance work and also the spare parts requirements in advance. This information also assists in analyzing the causes of poor performance of machinery due to environmental conditions. The results of failure analysis constitute a valuable feedback to the design and manufacturing departments that may attempt to develop advanced version of the equipment.

Training of Maintenance Personnel

The need of effective training of maintenance personnel has always been felt, but today it is more vital due to all-round induction of sophisticated/integrated equipment. For proper utilization of the maintenance workforce, the concept of total training of maintenance personnel is very useful. Even the persons from the production/service department must be trained so that they could understand the importance and need of the maintenance function. Therefore, training plays an important role in creating an atmosphere of congenial maintenance work without bias. Depending upon the need, special training programmes must be developed for all the concerned people of the organization. The job-oriented training programmes can also be more effective by giving due consideration to the basic qualification of the workers.

Before training is imparted to a person, an evaluation process should be carried out to know the type of training that is needed for the concerned worker. The management needs to decide the level of training that an individual should receive must be examined carefully. For the success of the training programme, self-study habits must be developed among the workers. The attitude of the maintenance personnel is very important before they are deputed for any training programme.

Development of Problem-Solving Groups

Maintenance personnel face several types of problems from time to time due the nature of the failure in the equipment. The formation of problem-solving groups helps to mutually solve the problems, especially where the fault is of a typical nature and involves association of diverse disciplines. Such groups can also do fault analysis effectively. Mathematical models and use of computers immensely help in the analysis of maintenance problems. The tailor-made expert systems designed for quick fault detection help to solve complex maintenance problems, especially in the area of diagnostics, troubleshooting, and failure analysis.

Life Assessment of Subassemblies

In view of the high cost of equipment and its accessories, efforts should be made to extend the life of the components/assemblies to the maximum possible extent without running into the risk of sudden catastrophic failure, which may cause serious safety hazards. Some suitable procedures should be evolved to make the life assessment of costly equipment/subassemblies. For example, the application of non-destructive testing (NDT) provides means for predicting the residual life of components and at the same time the vibration analysis of the support bearings will indicate their status and further use of the same.

Reduction in Maintenance Workforce

Due to limited availability of resources to any enterprise, the maintenance budget is generally axed and considered a liability on the production system. Efforts should therefore be made to reduce the maintenance workload with proper planning and scheduling. To keep the maintenance workforce as small as possible, the maintenance manager must know the answers to the following before planning:

- ◆ What constitutes the maintenance workload and how it is measured?
- ◆ In which areas can the workloads be reduced without affecting the plant operations, as well as the related maintenance work?
- ◆ How can labour productivity be measured and improved?

It is with better planning only, that the maintenance workforce can be reduced and this exercise should start with the assessment of the work and the skills of the workers.

The other areas of improvements include the use of dynamic scheduling for preventive maintenance programmes. In the case of fixed scheduling, the schedule is drawn for the entire year prior to the start of the year. Each piece of equipment is analyzed separately based on the frequency of all procedures to be performed and the time required for performing the procedures. The procedures are then assigned to appropriate weeks trying to keep the workload between the weeks as equal as possible.

Once the schedule generation process is complete, all procedures are tucked away in a nicely labelled schedule. Each week, the assigned procedures are assessed and performed. Some of the equipment may not get enough service. The easiest solution might be to alter the frequency of inspection for any specific equipment. Although this solution looks easy, it is difficult to implement the same in a fixed schedule, because all procedures need to be altered to the predetermined schedule. This process can be very difficult since the new procedures may not fit into the slots of the old procedures. Remember that the schedules should be kept as uniform as possible across all the weeks. To accomplish this task, procedure positions that were not originally affected may have to be altered. If the changes are large, the entire schedule might have to be redone.

Another problem area in fixed scheduling is the backlog of work. Once the schedule generation process is complete, work orders are generated each week considering the equipment to be serviced. If any piece of equipment is not serviced, it becomes backlogged. Since the schedule is prepared in advance, procedure to service the equipment must be worked out. When the equipment inventory is large, the backlog grows very fast. As the schedule is fixed, a new work order can be generated for a piece of equipment waiting in the backlog. As a result, some equipment may get serviced more than once within the stipulated time period.

In dynamic scheduling, there is no requirement for a previously determined schedule. Instead, the new work order schedule is fixed when the previous work had been done, not when it was supposed to have been done. Also, there are no fixed locations for procedures since no future scheduling is attempted. Because of this, any situation that causes rescheduling can be handled quite easily. All that is required is the changing of the procedure's frequency and the addition or deletion of procedures. Also, only those pieces of equipment that need to be changed are affected.

Backlogged work is also handled quite easily. The dynamic scheduling is aware of what happened in the past since it needs past information to determine what to do next. Therefore, the work that is backlogged is also considered in the scheduling process along with the work that is becoming due for maintenance.

The underlying factor in dynamic scheduling is to get the highest priority work completed as soon as possible. Therefore, in any given week, the workload will be quite high. However, only a predetermined number of hours of work are scheduled each week. If there is no work that can be scheduled, it is deferred until the following week. Therefore, no matter how high the pile of work is in front of the scheduler, it leaves a smooth trail of completed work behind.

It can be concluded that the fixed scheduling works well in a static and ideal environment. But since this type of environment is seldom found in practice, the dynamic scheduling, which is flexible, takes care of everything.

Equipment Maintenance/Replacement

To decide the effective mode of maintenance, it is essential to carry out reliability analysis of critical parts of equipment in all the modern automated and semi-automated plants. These critical parts may be individual pieces of equipment or a combination of parts that form a system.

Before considering the purchase of any capital equipment, the evaluation of its reliability is essential, which directly depends upon the probability of failures. It is desirable to obtain a reliability index (numerical value) for each machine, which is based on such factors as visual inspection, tests and measurements, age, environment and duty cycle of the equipment. These numbers, so calculated, represent the reliability of a particular equipment. The index must be determined for each piece of critical equipment. It is also possible to combine these indices and express an aggregate reliability index number for the complete system.

From the evaluation of the above index numbers, schedules can be set for equipment maintenance. Wherever needed, the maintenance efforts can be expanded. From the reliability reports it is possible to determine the actions that are required to maintain the operational availability at the desired level. Cost estimates for such maintenance functions can also be prepared, based on the reliability information.

Similarly, the decision to replace existing equipment will require the consideration of the following questions, economic factors and reliability index numbers calculated for the existing equipment.

(a) Will the maintenance cost come down with the replacement of the old equipment?

(b) Will the production cost per unit of production come down due to automated test features of the new equipment?

(c) Is the existing equipment not sufficient to meet the future production targets?

(d) Will the new equipment be environment friendly and provide better safety to men?

(e) Is there any possibility of adding additional accessories to existing equipment in order to make it more versatile for future use, or is the rebuilding of existing equipment possible through minor modifications?

By using the dynamic programming approach, the optimal replacement policy of the equipment can be determined if reliable estimates of revenue (return from equipment), upkeep (maintenance cost) and replacement costs are available.

For the equipment that deteriorates with time, the following model can be used to decide the replacement policy. The equipment in use in industries can be mainly divided into (1) equipment with diminishing efficiency and (2) equipment with constant efficiency. The first category deteriorates with time resulting in increase in operating cost including maintenance cost, and the second category operates at constant efficiency for a certain time period and then deteriorate suddenly.

Several models have been developed using repair versus time, and cost, in order to solve the replacement problem of equipment with diminishing efficiency. The concept of overhaul in lieu of replacement is one viable model as the operating cost increases with time. This model maximizes the gain between the operating costs before and after the overhauls. Overhauls are carried out to improve the equipment performance and thus reduce the operating costs. Replacement on the other hand is considered to be the regeneration point of whole life where the operating cost function initially starts. In practice, such methods really work well and the life of the equipment/system in enhanced.

INSTRUMENTED EQUIPMENT MONITORING

This system involves data capture, transmission processing and analysis of information and are used to address production analysis, operation cost analysis, production planning and maintenance control. In mining industry electric shovels have on board monitoring system which indicates, per cycle productivity, truck loading analysis, cycle time analysis, availability of equipment, operational efficiency, energy monitoring and control, mechanical and electrical diagnostics, predictive and preventive maintenance schedules. Similarly Draglines, Bucket Wheel Excavator, Haulage truck etc., used in open cast projects have the monitoring facilities as indicated above. For improving availability and reliability of all costly equipment/system the concept of instrumented equipment monitoring is being used. The usage of above mentioned systems indicate the following:

- ◆ Alarms under predetermined limiting conditions
- ◆ Feedback to motor control circuits, e.g. anti-tight line control
- ◆ Recording test data for troubleshooting or research
- ◆ Analysis of dig ability, including rates of bucket fill and specific dig energy per cycle
- ◆ With operator input, utilization delays and codes for particular activities are logged.

QUESTIONS

1. How do tribology practices help in improving the maintenance functions? Explain.

2. Specify the areas where terrotechnology practices can be applied effectively.

3. State how indigenous substitution can help in the case of imported equipment.

4. Reconditioning can also improve the performance of the equipment. Justify.

5. How does the maintenance information management system (MIMS) help in improving equipment performance? Explain.

6. Enumerate the areas where maintenance workforce can be minimized.

7. Briefly explain the concept of dynamic scheduling.

8. Compare the merits and demerits of equipment maintenance vis-a-vis replacement policy.

Chapter

10

Economic Aspects of Maintenance

INTRODUCTION

The optimum performance of machinery is a must for economic viability of any capital-intensive industry. The maintenance functions play a vital role towards achieving higher production targets, however, the cost of production operations must be within the laid down limits. To this end, every industry establishes a maintenance department to achieve their requirements effectively. One of the important areas of decision-making in the maintenance department is the economic consideration. The economic aspects of maintenance are looked into from different perspectives. The prevailing methodologies include the life cycle costing, value functions, performance-auditing, and the like. The economic analysis is also necessary for the management of inventory and adoption of the replacement policy.

LIFE CYCLE COSTING

It is the normal industry practice to purchase the equipment based on the lowest quoted price. However, some organizations may purchase higher-priced items on the grounds of superior performance and the like. Sometimes the cost of operation and maintenance is many times more than the acquisition cost of the equipment. This may be an indication of the drawbacks of the said equipment. The low acquisition cost of the equipment may be one of the reasons for its higher operating and maintenance costs. Therefore, it is important that the selection of equipment be based on the total ownership cost over its entire lifespan. This total cost of ownership is calculated considering the operations costs as well as the maintenance costs over the product's lifespan. Such type of costing is called the *life cycle costing*.

The concept of life cycle costing brings together engineering, economic and financial disciplines in connection with the equipment to be procured. The economic life of equipment depends on its maintenance and repair costs, and the resulting availability and operational efficiency. It can be estimated by plotting the cumulative efficiency and maintenance and repair cost per cumulative hour against operating hours. The interaction of the two curves will indicate the economic life (Figure 10.1).

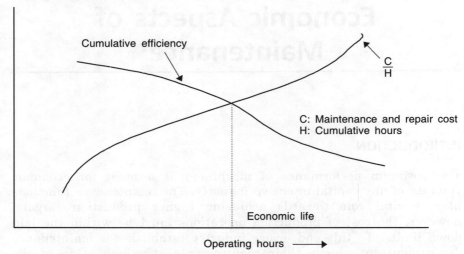

Figure 10.1 Estimation of economic life of equipment.

The aim of life cycle costing is to know the total cost of equipment accruing over its whole life period, which may include all the costs starting from the specification cost. It is also observed that reduction in one cost may increase the other cost. The total cost has to be therefore optimized by making trade-offs among all elements of it. Normally it is observed that cheap items do not have sufficient economic life nor do they yield satisfactory performance over their whole life. Therefore the quality of the product is of prime importance.

The various elements of cost involved in ownership of an equipment are as follows:

◆ Acquisition cost (at the time of purchase inclusive of all allied costs).
◆ Installation cost including commissioning and testing costs.
◆ Spare parts cost including the cost of all support materials such as test equipment, jigs and fixtures, and the like.
◆ Operating cost of the equipment including the cost of facilities, utilities, operating personnel and their training.
◆ Maintenance cost including the cost of training of maintenance personnel.

The main advantages of the life cycle costing are as follows:

◆ It may result in selection of an equipment that has lower operating and maintenance costs resulting in reduced cost of ownership.
◆ The money saved can be used for some other works.

MAINTENANCE COST

During the planning stage, budgets are normally allocated for each type of activity, which must also include the maintenance cost. In the case of production the cost can be set easily. Whereas, in the case of maintenance function, it is difficult, since failures are random phenomena and their levels cannot also be defined. The usual practice to set targets for maintenance cost is based on the past experience. The maintenance cost, however, will directly depend on the level of maintenance and its requirements. It has been estimated that about 70 per cent of the cost of maintenance, spare parts and materials is incurred on account of replacement of components and assemblies subjected to wear and tear, and 25 per cent of this cost can be saved by a systematic application of advanced tribology practices. Such improvements enhance the useful life of equipment by containing the rate of wear. Consequently there is reduction in downtime of the equipment. It enhances the production as well. As a result, the demand for spare parts reduces and it becomes possible to manage the maintenance work with less manpower. All these factors have a great potential towards contributing to overall reduction in the costs of the products as well.

A study conducted in a steel plant showed that the profitability may be increased by improving the maintenance-engineering practices. This is possible through advanced maintenance practices. Table 10.1 indicates the gains realized through better maintenance practices for some selected items used in the maintenance function.

Table 10.1 Ways of cost reduction in steel plants through better maintenance practices

Major heads of maintenance cost	Current cost as % of production cost	Cost reduction by 25% improvement of direct material cost
1. Spares and material for regular maintenance	4.75	3.56
2. Lubricants	0.95	0.89
3. Spares and subassemblies for long-term maintenance	1.30	1.22
4. Maintenance management and labour	3.0	2.81
(A) Direct maintenance cost (1 + 2 + 3 = A)	10	8.48
(B) Cost of downtime (2 × A)	20	16.96
Total maintenance cost (per cent of cost of finished steel) inclusive of cost of lost opportunity (A + B)	30	25.44
Potential saving		4.70

The maintenance cost can be estimated by knowing the following:

◆ Number of breakdowns with their levels.
◆ Downtimes of the equipment for want of repairs.
◆ Consumption of spare parts, equipment-wise.
◆ Penalty cost, if any, due to non-availability of equipment for production work.

The task of estimating the maintenance cost is difficult, since it depends on the frequency of failures and efficiency of maintenance services. Operating personnel are responsible for equipment usage but the maintenance personnel shoulder the responsibility for the maintenance function. The maintenance manager has to deal with this dual problem tactfully for an effective planning and proper control of maintenance cost. The analysis of maintenance cost is very essential prior to making any decision regarding replacement of a machine or any of its components.

IMPACT OF MAINTENANCE COST

The cost of maintenance may differ from one organization to another depending upon the importance attached to the maintenance function. It is seen that for critical equipment the cost of maintenance is normally high because in such cases, it is the preventive maintenance that is followed. The maintenance cost includes the cost of spares, materials, personnel and expenditure incurred on services such as electricity, water, air, gas, etc.

The other dimension of maintenance cost is the downtime of the equipment. In real life, the downtime cost can be much higher than the actual maintenance cost. It may be even double the cost of maintenance for some specific type of equipment/services.

The demand for the product being manufactured by the company plays an important role insofar as the downtime cost is concerned. To maximize the production, the availability of the equipment has to be high. The downtime of the equipment can be minimized through planned maintenance and by increasing the life of the components and subassemblies of the equipment. This can also be achieved by monitoring the condition of the equipment at appropriate time intervals. An improved component life means less frequent replacements and therefore reduction in workforce for such jobs. All these factors help to minimize the overall cost of the maintenance function. A planned maintenance system can provide the maintenance function at minimum cost.

The maintenance cost is comprised of two factors — the *fixed cost* and the *variable cost*. The fixed cost includes the cost of support facilities including the maintenance staff, and the variable cost accounts for the consumption of spare parts and the use of other facilities as required during the time of fulfilling the maintenance requirements. Thus it is seen that the variable maintenance cost has direct relationship with the maintenance function and reducing the maintenance activities or utilizing the spare parts of improved quality can thus minimize component cost.

The maintenance cost does not have any direct relationship with the production cost. It may be minimal during peak production months and very high during low production periods. This may be because of management's deliberate policy to defer the maintenance activity either due to fund constraints or due to production requirements. Even painting and denting may help in increasing the life of the equipment, but these are normally deferred due to lack of funds or production targets.

MAINTENANCE BUDGET

For effective working of the maintenance department it is essential to set aside certain amount of money to meet the maintenance expenditure, which is termed maintenance budget. The maintenance budget can be of the following types:

- ◆ Appropriation of maintenance budget, which sets aside money for each activity independently.
- ◆ Fixed budget, which means fixed allocation is made for a specified period of time for overall maintenance functions.
- ◆ Variable budget, which is based on the requirements of the maintenance functions and can take care of all exigencies.

For efficient working of the maintenance department, it is desirable that all of the above types of budgets be judiciously prepared. The biggest challenge in preparing such budgets is the unpredictability of the maintenance jobs. It is really difficult to forecast the cost of breakdowns as the level of such breakdowns itself is not known. The position of such breakdowns is also important looking into the inconveniences caused to other users.

For the purpose of cost control, it is relatively easy to implement material and time standards for production operations. However, the same is not possible in the case of the maintenance function. The decisions to be taken in the case of the maintenance function depend upon various factors. To aid decision-making, comparisons can be made with various existing systems after obtaining the cost data.

The factors described in the following subsections should be considered during the process of decision-making in maintenance activities.

Preventive Maintenance versus Breakdown Maintenance

The relevant information regarding the failure data, the downtime cost arising from breakdowns, cost of damage, loss of production, penalty cost and cost of replacement of damaged parts, and so forth should be analyzed in respect of the above systems to compare the nature of maintenance to be adopted under a given situation.

Replacement versus Repair

If the cost of maintenance as well as that of operation shows an increasing trend, it is desirable to replace the affected equipment/part. Here, the benefits from the replacement must be known to adopt a particular policy, however, the cost of equipment/service must be kept in mind. Replacement also allows procurement of a better quality equipment, though it may be more costly than the existing one. It should be the general concern of the management to take care of deterioration and obsolescence.

Hired versus Departmental Services

The variable as well as the fixed elements of costs of the maintenance department can be collected and a comparison made when the same facilities are hired from outside. In the present context it is seen that the hiring of the services can be more effective and economical in the case of public enterprises because of high fixed costs that include manpower and material costs.

Overhaul Periods

Schedules can be prepared to repair the equipment during its normal downtime in order to reduce the production costs as well as the machine downtime costs. Here also, the relevant costs such as the cost of production, hiring charges and downtime cost can be collected for the purpose of comparison. However, the overhauls if carried out in time will improve the performance of equipment and enhance its service life.

Maintenance-In-Time versus Time-Availability Maintenance

In some places, it is a normal practice to run partially-failed equipment for the purpose of meeting the production targets. This may deteriorate the condition of the equipment further. Information regarding consequential loss of efficiency, damage to equipment, etc. can be collected for a better decision. However, such practices must be discouraged in order to achieve the satisfactory performance of equipment/services.

It is possible to take a decision about a particular policy, if the cost of relevant maintenance and related functions are known in advance so that better maintenance cost control can be exercised. In actual practice the maintenance-in-time can yield better results as compared to time-availability maintenance.

Cost of Obsolescence

To match the productivity of newly available machines, the older machines are often subjected to excessive stress. There are always higher expenses associated with fuel, oil, wages, and maintenance of older machines. Such expenses are termed cost of obsolescence. Sometimes running of obsolete machines causes excessive downtime due to nonavailability of essential spares in time.

COST CONTROL

The maintenance department has to exercise effective cost control to carry out the maintenance functions in a pre-specified budget, which is possible only through the following measures:

- ◆ First line supervisors must be apprised of the cost information of the various materials so that the objective of the management can be met without extra expenditure on maintenance functions.
- ◆ A monthly review of the budget provisions and expenditures actually incurred in respect of each centre/shop will provide guidelines to the departmental head to exercise better cost control.
- ◆ The total expenditure to be incurred can be uniformly spread over the year for better budgetary control. However, the same may not be true in all the cases particularly where overhauling of equipment has to be carried out due to unforeseen breakdowns.
- ◆ The controllable elements of cost such as manpower cost and material cost can be discussed with the concerned personnel, which may help in reducing the total cost of maintenance.

It is observed through studies that the manpower cost is normally fixed, but the same may increase due to overtime cost. However the material cost, which is the prime factor in maintenance cost, can be reduced by timely inspections designed to detect failures. If the inspection is carried out as per schedule, the total failure of parts may be avoided, which otherwise would increase the maintenance cost. The proper handling of the equipment by the operators also reduces the frequency of repair and material requirements. Operators, who check their equipment regularly and use it within the operating limits, can help avoid many unwanted repairs. In the same way a good record of equipment failures/ maintenance would indicate the nature of failures which can then be corrected even permanently.

A timely submission of equipment for service/repair by the production personnel can contribute towards reducing the rate of material consumption, because it would avoid further deterioration of the parts which can be repaired and therefore do not require to be replaced. The objective of cost control is to achieve savings. For this purpose, estimations of economic life, total cost of owning, running, and maintaining a machine are essential. As shown in Figure 10.2, if the average life cycle cost of a new machine is known and the combined cost of operating and maintaining an old machine is evaluated, the exact time for replacement can be decided and savings of some expenses can be obtained.

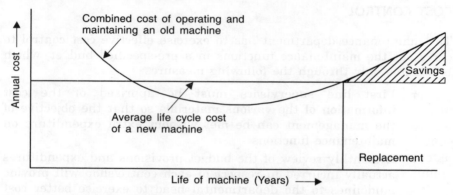

Figure 10.2   Cost-based replacement of machines.

ABSENTEEISM AND OVERTIME

One important factor leading to higher maintenance cost is the cost due to absenteeism and overtime. Because of fixed manpower provided for the maintenance work, the cost of labour is influenced by absenteeism of the labour and overtime payments made to them. Even though workers may be absent due to valid reasons such as medical leave, earned leave, etc. being regular employees, other persons may have to be hired on an ad hoc basis and this ultimately increases the cost of maintenance. Sometimes the regular employees are paid overtime, thus increasing the maintenance budget. To minimize such extra payments, the maintenance department needs to follow strict policies to combat absenteeism and overtime. A study is conducted to frame a policy concerning composition and size of the maintenance workforce. The workload should be measured. This will help to ensure the smallest and most effective workforce. By knowing the workload, the number of men required to complete a job can be determined. For every group of maintenance workforce, the targets must be set and then the actual work carried out by the group should be measured. Workload can be measured by dividing the total maintenance work into categories that lend themselves to easy measurements. For example, in planned maintenance the nature of work is known in advance and therefore it is easy to determine the manpower requirements for the same.

Productivity is a measure of the effectiveness of labour control and in maintenance it is the percentage of time that workers perform productive work. Therefore, if the maintenance work is viewed as productive, it will help in reducing the maintenance costs. For this, the performance measurement can be designed in the form of reports to control the workforce. An effective use of available manpower can also contribute towards minimizing the maintenance costs.

The studies conducted on maintenance function show that the labour cost can be a good measure of labour efficiency. It has further been observed that by upgrading the planned maintenance activities through proper planning and scheduling it is possible to substantially reduce the labour cost.

Effective planning of maintenance work is one of the most important ways to minimize the maintenance costs. Planning ensures that material, labour, tools, work place, etc. are properly arranged and utilized. This helps the maintenance crew to complete the job in a well-organized and productive manner. It is seen in practice that maintenance personnel in laying their hands on the above facilities waste a good amount of time, and such infructuous effort can be minimized through better planning and control. The supervision of maintenance work is also very essential to control the costs. Many times the work is undertaken without proper approval and planning, which not only results in improper use of the maintenance workforce, but also leads to improper distribution of maintenance workload.

MAINTENANCE AUDIT

A maintenance department is the centre of any organization and therefore its audit is of immense help to the management to bring down the cost of maintenance related activities where its proportion is high. In a drive to reduce maintenance cost the management must identify the real cost drivers and must evaluate the risks of any cost reduction measures like reduction of manpower, spare parts or training etc. Some important points justifying maintenance audit are:

It may be necessary to understand the total investment needs for modernization and performance improvement for a particular equipment/system and may help to know the following:

- ◆ Requirement of maintenance tools and equipment
- ◆ Type of maintenance system

- Maintenance organizational structure
- Maintenance training for the workers and supervisors
- Impact of maintenance strategy

Maintenance information obtained through audit will give idea about scope of return on investment, which can be considered for improvement.

This can also indicate cost benefit analysis or risks associated with outsourcing maintenance jobs to maintenance experts.

It helps in developing performance indicators to incorporate improvements.

It can also help in deciding replacement policy of the equipment/system with advanced technology. Where similar projects are established at various locations maintenance audit will help in standardization of maintenance practices and other resources required.

It will also help in improving plant performance with better profit margins.

Decision making in context with maintenance related problems becomes much easier when sufficient and accurate information is made available to the planners.

Maintenance Audit Procedure

Before starting the maintenance audit the methodology for it must be well defined to achieve the desired results. The answers to the following questions must be kept in mind before implementation.

- What are the requirements of maintenance audit?
- Whose responsibility will it be and what skills will be needed for it?
- What and how much will be audited?
- What audit approaches and methodologies will be used?

It should be worn in mind that maintenance audit and inspection are different. Inspection during maintenance identifies the maintenance work requirements, whereas audit provides detailed information about fulfillment of rules and regulations, safety and maintenance objectives.

To carry out maintenance audit a team of qualified and experienced persons has to be formed under a leader who will supervise the work. The team leader has to ensure that the audit work is done systematically to get the meaningfull results. Depending upon the audit need, it could be full audit, snapshot audit or fingerprint audit, which involves the following:

- *Information collection:* For the purpose, interviews of key persons must be conducted including suppliers and contractors. From this questionnaire and survey forms can be designed.

- Conducting site inspections of work centres, service facilities and associated buildings with full information for the establishment of maintenance department.
- *Data analysis:* For identification of problem areas and development of improved organization and systems.
- Laying process flows and mapping maintenance functions and control with review.
- Development of store management for spares and facilities.
- Identification of systems and subsystems with their utility.
- Attending decision-making meeting.
- Involvement in all maintenance functions.
- Presentation of correct performances of plant, equipment and maintenance functions.
- Preparation and presentation of audit reports.

Audit Tools

These are the means of analyzing the maintenance functions and controls, which are developed by the maintenance team according to the requirements of the auditing industry.

Method of auditing involves the following:

- Determination of efficiency of each and every component of the system
- Collection of evidences
- Conducting surveys with proper pre-survey studies
- Consultation with fellow members
- Conducting interviews with management
- Carrying out reviews of authorities, policies, directives etc.
- Undertaking thorough review of performance reports
- Scrutiny of results
- Observation of available facilities in detail
- Identifying and understanding of all the major systems and control procedures
- Analyzing the relationship between resource utilization and results obtained
- Assessment of associated risks to find out:
  - What can go wrong?
  - What is the probability of going wrong?
  - What are the consequences?
  - Can the risk be minimized or controlled?
- Consultation with advisors and outside organizations to identify the best practices and opportunities to improve
- Exploring previous studies and audits conducted, if any
- Technology selection through survey
- Thorough review of the trend available from the studies

Audit Planning

Planning is required for effective conduct of maintenance audit with the emphasis, what to audit, since auditing each and every thing will be time consuming and costly. Therefore, it is needless to point out that selection of audit areas must be made judiciously. During the planning stage one must identify the important factors that affect maintenance functions and assess how these factors are responding to key challenges, opportunities and critical success factors. The following must be considered:

◆ Examine whether the areas selected can be audited properly which can be done with the help of experienced audit team.

◆ Develop audit schedule for the implementation of audit policies.

◆ Quality control of work under consideration by a qualified and experienced reviewer.

◆ Plan to obtain views on critical success factors for the activity being audited, management responsibility for the activity, risks, management concerns etc.

◆ Setting up of Audit Advisory Committee with experts in all related fields.

◆ Preparation of documents for audit work with proper files and evidences.

The organizational structure is very important for the audit, as duties of the personnel attached with the work are crucial for the success of the programme. Management responsibility is aligned with the value chain of the business and represents the who, what, why and how in order to successfully maintain the targeted production quantity and quality. Most importantly, management determines the business reasons for maintenance.

Systems and procedures are the tools, either imported or developed within the organization, to effectively administer management's visions and requirements and provide best practice assistance to the workforce. Personnel and resources carry out the assigned work to the expectations of the management. The links between these key areas are the drivers of organizational culture and directly represent management's commitment and leadership.

A thorough maintenance audit has to question all these areas and links in order to present a total understanding of the business and to match the concerned plans. The audit evaluation table is formatted to provide an objective assessment of each element. The purpose of this is to provide the auditor with an understanding of the status of the maintenance function development within the organization.

The audit summary sheets are normally used to prepare short/ long term improvement programmes focused on recommendations and requirements.

Audit Reports

The outcome of the audit is presented in the form of reports, which may include the following:

- ◆ Preliminary audit objectives
- ◆ Overview and background of the functions under audit
- ◆ Issues considered and reasons for their acceptance
- ◆ Survey findings
- ◆ Proposed audit scope
- ◆ Draft audit criteria
- ◆ Timings for events and meetings
- ◆ Audit approach
- ◆ Auditing results directly, e.g. cost effectiveness of the hired maintenance persons
- ◆ Auditing the control systems, e.g. cost effectiveness of the maintenance breakdown reporting system
- ◆ Flow charts to analyze system

All must present the results of the maintenance audit to the management in a manner such that it is easily understood. The following formats can be used:

1. Management summary and essential recommendations
2. Audit evaluation table for each element
3. Audit summary sheet for each principal function
4. Audit comments and recommendations for each element

Besides above the audit report must include improvement plan for implementation.

It is observed that the proper conductance of maintenance audit will provide sufficient information about the financial status of the organization with suggestions to improve the activities wherever needed. Maintenance audit should be considered as a must for proactive decision making because cost-effective improvement can be accomplished only when management understands the effectiveness of the maintenance organization. It is assumed that the maintenance work burdens the production targets.

QUESTIONS

1. Briefly explain the concept of life cycle costing of equipment.
2. Indicate the main components of cost involved in ownership of an equipment.

3. How does maintenance cost affect the production cost? Explain.

4. Discuss the impact of maintenance cost on the overall cost of production.

5. Briefly explain the factors, which play an important role in the preparation of maintenance budget.

6. Explain how cost control in maintenance function can be effectively applied.

7. Discuss why it is essential to consider the economic aspects of the maintenance function.

8. A residential campus is planning to procure a standby generator. Discuss how will you estimate the ownership cost for the whole life cycle of the generator.

9. How would you determine the exact time of replacement of a machine? Develop a decision table for the replacement of machinery.

10. Derive an expression for the calculation of downtime cost of equipment. For a computerized maintenance management system, how can you incorporate online maintenance costing service for the planners and decision-makers. Design some suitable user interface to use a maintenance database for cost queries.

11. What do you mean by maintenance budgeting? Develop guidelines for maintenance budgeting for

    (a) a pump house,
    (b) a transportation company with 40 trucks, and
    (c) a plant with 10 presses.

12. Explain how maintenance audit helps in improving the production of any organization.

13. Briefly discuss the audit procedure giving important points.

Chapter

11

Organizational Structure for Maintenance

OBJECTIVES OF MAINTENANCE ORGANIZATION

Maintenance of modern machinery and industry calls for setting up of a healthy, balanced and rationalized organization to achieve the multi-faceted objectives. This is essential to keep the diverse activities of the maintenance department under control without any problem. An effective maintenance organization can be set up after a careful study of the background of the industry because the nature of machinery deployed would be the key factor. It is seen in practice that some of the equipment need more attention than the others and the effect of environment is also important for the establishment of maintenance department. While setting up the organization, the future trend of the industry should also be assessed so that any modifications required in the set-up can be incorporated in due course of time.

The importance of the maintenance organization grows with the increasing size of the industry and the complexity of the equipment/ system. A small-scale industry may require a simple maintenance organization with a foreman in-charge. Whereas, when the size of the industry increases, the complexities of the maintenance and therefore those of the maintenance operations, also grow in size. And therefore, a highly trained and competent person may be required to head the maintenance organization/department. There has to be a close relationship between the production and maintenance departments for their effective working without problems.

The basic objective of any maintenance department/organization should be to ensure that the maintenance function is carried out effectively and the equipment downtime as well as the loss of production is minimized. High equipment availability can be achieved by proper coordination of maintenance functions with other departments. At the same time the cost of maintenance functions must be kept under control since it is considered a burden on the production.

The head of the maintenance department/organization should possess the necessary expertise and experience since the success or failure of the maintenance department would depend upon his abilities and capabilities. He has to look after not only the day-to-day maintenance/repair of equipment but also ensure the lifeline of the production unit or process unit. High availability of the production system can be obtained by continuously monitoring the planned preventive maintenance schedules identified for the machine/equipment. In a highly mechanized and sophisticated organization, the chief engineer (Maintenance) should normally head the maintenance department. The structure of the maintenance department/organization depends on the type of the industry as well. For a complex industry, for example, mining, a centralized maintenance unit can be set up since hundreds of mining units operate independently in different areas. In such type of organizations, a maintenance unit can be set up at the headquarters, and several field branches of such a unit can be set up at convenient places near the working areas to cope up with the daily requirements.

The following points should be considered while setting up a maintenance organization or department.

◆ To organizational goals pertaining to the maintenance function must be identified and should match with the overall organizational goals.

◆ Maintenance workload should be determined beforehand to achieve the desired results with proper utilization of manpower and facilities.

◆ Total maintenance work should be properly and uniformly distributed among all the concerned persons to avoid conflicts among the workers.

◆ The important and essential work should be assigned clearly and definitely to various departments of the maintenance organization.

◆ Due consideration and care should be given for proper qualification of the workers while assigning a particular job.

◆ The maintenance organization should be staffed with the best-qualified and trained people available since the nature of work is unpredictable and time consuming.

◆ The policies and procedures should be clearly designed at an early stage to help maintenance people to achieve the organizational goals.

In developing a maintenance organization, it is necessary to achieve a compromise between the individual needs and the organizational goals.

MAINTENANCE FUNCTIONS AND ACTIVITIES

A good maintenance organization should reflect the total maintenance activities required to be undertaken by the various sections of the department. The functions and activities associated with the maintenance organization are as listed below:

◆ Identification of areas/components/systems for implementation of the preventive maintenance, or other maintenance pro-grammes as applicable.

◆ Arrangements for maintenance facilities for carrying out the maintenance work properly and promptly.

◆ Planning and scheduling the total maintenance work of the organization keeping in mind the future and emergency requirements which are likely to come in due course of time.

◆ Ensuring proper and timely supply of spare parts through an efficiently managed inventory control system. In this case use of computers can be made with well-established inventory models.

◆ Carrying out standardization of maintenance work to help the organizations using similar type of equipment/system.

◆ Arranging import substitution of components, if necessary.

◆ Implementation of modifications, to the existing old equipment for improving performance, wherever possible particularly in case of imported items.

◆ Assisting the purchase department in procuring new machinery, construction work related to machines, and replacement of the parts/equipment.

◆ Disbursement of services such as, water, compressed air, steam, electricity, and the likes required for maintenance functions.

◆ Identification of obsolete, non-performing and surplus equipment/system for replacement, re-modification and disposal.

◆ Training of maintenance personnel on latest know-how in relevant areas.

◆ Ensuring safety of persons, as well as equipment/system in the organization.

ORGANIZATIONAL REQUIREMENTS

While setting up a new maintenance organization/department, the following parameters must be kept in mind.

Coordination

Success of an organization depends on effective coordination and cooperation of the people in the organization. For a prolific maintenance organization it is essential to have a clearly defined organizational chart indicating job-descriptions and unambiguous definition of authority for each job. It is observed that a group rather than an individual performs most of the maintenance jobs. Therefore, it is utmost necessary to maintain work harmony. To maintain proper team spirit, the leadership of the group is very important which can bring people together. Thus, success of the maintenance function entirely depends upon the dynamic leadership and coordination of the departmental head.

Maintenance Engineer

The role of the maintenance engineer is very important and crucial, as his primary job is to minimize the downtime with optimal performance of the equipment/system under his supervision and control. Therefore, the maintenance engineer should be a well qualified, technically competent, knowledgeable and cost conscious person in order to be an effective and efficient leader of the maintenance group. The output of the maintenance department with respect to time, talent and effort put in by the team is the index of its efficiency. Application of proper management techniques, motivation of workforce, avoidance of conflicts and the inspiring leadership of the maintenance engineer can bring the targets and achievements closer. Accountability of each and every individual of the maintenance group is essential for effective utilization of the manpower. The authority wrested with the maintenance engineer should not be abused in the interest of the organization.

Attitude of Personnel

The attitude of the working personnel in many instances is not always positive towards accepting the modern facilities/techniques available for the maintenance function. Under the existing conditions the maintenance personnel should be apprised of the latest techniques available in the field of maintenance together with their merits and demerits so as to change their mindset and encourage them to adjust to the changing environment. The organizations must correspondingly mould their policy to bring change in the attitude of their personnel towards acceptance and implementation of advanced maintenance concepts and techniques, which can definitely help to improve the performance of maintenance personnel. The persons in the organization must be encouraged and motivated to accept the challenging jobs of maintenance which otherwise is taken as secondary job.

Policy

The policies adopted by the maintenance organization should be such that they can be easily understood and followed by everyone within its framework. These can be either formal or informal. The formal policies are recorded and are subjected to modification from time to time. It is the responsibility of the maintenance boss to frame the policies, in consonance with the company rules, keeping in mind working hours, benefits, safety, security, overtime, and the like. The policies framed should fit in the organization set-up as well. On the other hand, informal policies depend upon the situation of the maintenance requirements and are to be decoded on the spot. This situation normally arises in case of sudden breakdowns and no planning on the spot is possible. And therefore, such decisions may bring criticism from the fellow members as well as production people and require due attention.

Maintenance Work Control

The structure of a maintenance organization should be such that there is always sufficient manpower to supervise the work. Care should also be taken to see that a balance is always struck between the number of supervisors and the number of workers to be supervised. However, there is no fixed rule to maintain the proper ratio of supervisors to workers, although effective monitoring is most essential because, when the number of supervisors is high the work may suffer at the same time their shortage will also yield the same result.

Managerial Functions

It is the duty of the supervisory staff to develop their subordinates to a required level of proficiency. The working personnel should be properly motivated to attain the high levels of proficiency so that they can work freely and fearlessly to fulfill the organizational objectives. However, the attitude of the working personnel is very important for the development of the subordinates. The development of the subordinates can be done keeping the following in mind:

- ◆ Level of education and training
- ◆ Level of skill
- ◆ Level of intelligence
- ◆ Level of motivation
- ◆ Attitude towards the work

It must be pointed out here that the above levels entirely differ from one person to another. The jobs to be handled also require different levels of skill, such as unskilled, semi-skilled, and highly skilled however, the importance of experience will have its own weightage. This should also be borne in mind during the development stage of the subordinates. In this regard, the leadership of the supervisor plays a very important role to coordinate the work effectively.

Capacity Utilization

The main function of the maintenance organization should be to utilize the full potential of its workers and use of maintenance facilities. During the selection of persons, the values of each individual must be assessed thoroughly, as it is seen that though a person is a very good engineer but lack managerial qualities. However, managerial qualities can be imparted with proper training and incentive and thus a good engineer can become an efficient manager as well. It is also sometimes noticed that individuals move from one organization to another for better career opportunities. This can be taken care of by the top management so that the skilled and efficient persons are retained in the organization. The real leadership calls for an effective utilization of the capacity of the individuals with utmost care. This can be best achieved by properly assigning the jobs to the personnel. The maintenance work should be projected as a challenging job so that the persons, who are competent and dynamic, do not hesitate to take up this assignment.

ORGANIZATIONAL PROBLEMS

The maintenance organizations face various problems due to the nature of the maintenance work as well as its occurrence place. To overcome these problems, the maintenance organizations, with the cooperation of the plant management and the production department, should try to fill up the organizational gaps arising out of the following situations.

Lack of Coordination

The plant management must identify the objectives of the maintenance organization in consonance with the overall production strategy. Sometimes the production department does not allow the maintenance personnel to take up the scheduled maintenance work as this disturbs their production targets. This normally happens during the end of the production month.

The maintenance work is of varied nature; the maintenance organization should understand the difference between the routine and essential work so that the essential work is carried out most expeditiously with the cooperation of the production department and the routine work can be adjusted during other available time period.

A clear distinction between field and shop work should be made so that the traditional attitude does not hamper the maintenance work. The logic will not hold good for the breakdown maintenance work all the times because of the nature of failures. The maintenance organization should try the new concepts for better work control procedures.

Availability of Trained Manpower

As stated earlier, the properly qualified and experienced persons should only man the maintenance organization. Sometimes the persons from other departments are also shifted to the maintenance department either on promotion or due to their unsuitability in their respective departments, which should be avoided as it may create serious problems in the maintenance organization.

It should be borne in mind that though policies set up by the top management should act as the guidelines for every activity of maintenance function, the actual procedures are the outgrowth of such policies because of the nature of work involved.

It is normally seen that if more than the required number of people are engaged on a job, it ultimately suffers, as nobody owns its responsibility. Sometimes, too many assistants in a particular job create confusion among the workers due to multiple chains of command, which must be avoided to minimize conflicts among the workers.

Improper Planning

During the planning stage of an organizational set-up, it must also be kept in mind that overstaffing needs to be avoided to obviate overlapping of duties and at the same time understaffing will also result in not completing work as per schedule therefore proper balance for all type of activities must be struck. The planning of other activities associated with maintenance functions is also important and crucial and must be cared well in time. This include the facilities required for completion of any maintenance related problems.

TYPES OF MAINTENANCE ORGANIZATIONS

The selection of a particular type of maintenance organization/system will largely depend upon the main structure of an industry as well as the type of product handled. However maintenance organizations in general can be of the following types:

- ◆ Decentralized
- ◆ Centralized
- ◆ Partially decentralized

In the large-sized plants located at different places, inter-unit communication is difficult. Therefore, in such cases the decentralized type of organization is best suited, which means that the maintenance organization works under the direct control of a chief engineer in-charge of production. With this type of organizational structure, a better coordination between the production and maintenance groups is possible since one person heads it only and the communication down the line is direct. The advantages of such type of organization are as under:

- ◆ Speedy and timely decisions due to better line of communication under single control is possible.
- ◆ Maintenance and production people understand each other's problems better because of their common goals of meeting the production targets.
- ◆ Interchangeability of workforce, even at the managerial level, is also possible without any conflict among the staff.
- ◆ Better training at the workers' level can be arranged.
- ◆ Better coordination with fellow members.

In a small factory where communication between the departments is more free, the centralized type of maintenance organization is preferred, which is placed under a chief maintenance engineer/manager. The total responsibility of the maintenance function for the entire factory lies with the chief maintenance engineer. Under this type of organization, the responsibilities and accountability of work must be properly specified for production as well as maintenance personnel to successfully meet the project goals. If this is not taken care of, one department may blame the other for any shortfall.

The partially decentralized organization, which is the modified form of a centralized type of organization, is most suitable for projects that have units at far away locations. Under this type of maintenance organization, the maintenance personnel, attached to the production unit, carry out work at unit level and look after day-to-day maintenance. However, important maintenance functions such as overhauling, planned maintenance work, procurement of spare parts, etc. is kept under the charge of a chief maintenance engineer at the central office. All the centralized work pertaining to maintenance planning and documentation is done at the level of central maintenance workshop.

The above three types of maintenance organizations, however, are not strictly exclusive and some adjustments can be made to suit the working environment and the need.

Basically there are two types of organizations that are followed in most of the projects/industries—line organization and staff organization. In the past, the line organization was mostly followed, which consisted of a general foreman and a number of specialist foremen with their own working personnel under them. Here the specialist foreman executed maintenance work in their respective areas while the general foreman supervised the total work under his control. Figure 11.1 shows a typical line organization.

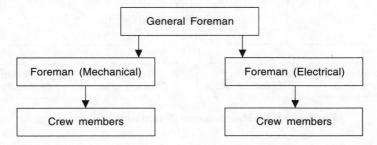

Figure 11.1 Line organization.

Realizing the need of recording the maintenance work, a few staff members when introduced in the line organization form a line-staff organization (Figure 11.2).

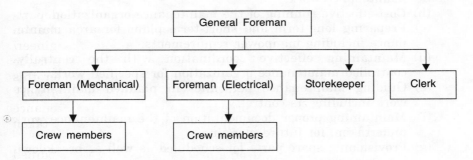

Figure 11.2 Line-staff organization.

In the recent past, the maintenance organization was based on craft concept. The persons joined the organization as apprentices and rose to the higher positions (foreman) after accumulating sufficient experience in their jobs. However, these persons were never given a formal training of the foreman's job. Such an organization emphasized craft skills and therefore strongly supported the idea of specialization gained through skills. Further growth in the field of maintenance function led to the concept of functional organization that was purely based on craft organization. In this system a few workers are placed under each functional foreman (Figure 11.3).

In the mining industry, the area type of maintenance organization provides better utility of available manpower. Here a single person takes care of maintenance for a particular given area in a well-defined manner. He is provided with a number of workers to carry out day-to-day work including emergency work and to coordinate with centrally controlled shops for the other important maintenance work.

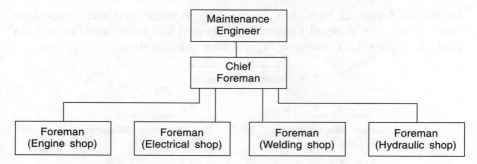

Figure 11.3   Maintenance functional organization.

Because of less amount of maintenance work at the area level, the utility of manpower is usually low but with better interaction with the central workshop, this problem can be eliminated to some extent. The duties and functions of an area maintenance engineer are listed below:

(a) Establishment of well-defined standards and practices for the maintenance work.
(b) Cost-effective running of the maintenance organization.
(c) Preparing long-term and short-term plans for area maintenance including manpower requirements.
(d) Maintaining effective coordination with the centrally-controlled maintenance organization for specified work.
(e) Guiding maintenance personnel in planned maintenance work including its control.
(f) Maintaining proper documentation of the maintenance work undertaken, for future planning.
(g) Provisioning spare parts for scheduled as well as breakdown maintenance.

The area maintenance organization, as shown in Figure 11.4, may be adopted for engineering services.

In some of the industrial organizations, the centrally controlled type of maintenance systems are adopted where all major types of maintenance work are undertaken by the central workshop. This is generally possible for all types of equipment, which can be shifted to the workshop. Moreover, such type of maintenance work can be taken up during the shutdown period of the plant. The maintenance superintendent, as indicated in Figure 11.5, controls this type of organization.

Some subjective type of analysis can be done to find out the merits and demerits of different types of maintenance organizations, for the final selection of one for a particular project/industry. For example, the production superintendent may also be in-charge of exercising the control over the maintenance organization. But in big projects, a separate person may be doing this work independently. It is sometimes seen that the production superintendent utilizes the

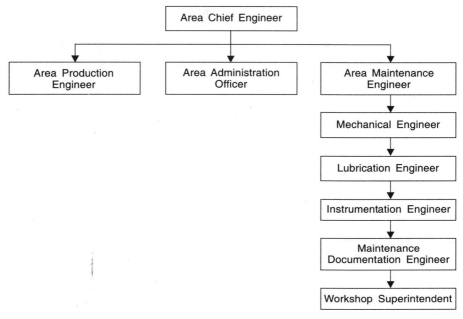

Figure 11.4 Area maintenance organization.

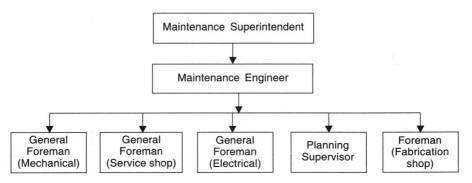

Figure 11.5 Centrally-controlled maintenance organization.

equipment longer than the scheduled time period to meet the production targets, which ultimately affects the service life of the equipment. This type of situation can be avoided if two independent persons are made responsible for the two different jobs. The operators are also sometimes made responsible for the first-level maintenance function. Such an arrangement is practised in many industries and has been found to be feasible and successful.

COST MINIMIZATION IN MAINTENANCE ORGANIZATIONS

Containing the overhead costs in maintenance organizations calls for budgetary controls. Adequate steps in the following directions can lead to the minimization of costs.

- ◆ Centralized planning, scheduling, and controlling
- ◆ Grouping of specialized workforce
- ◆ Effective labour utilization
- ◆ Budgetary control
- ◆ Use of contract maintenance to reduce the overhead costs on manpower and equipment in the areas where expertise is required and difficult to maintain the equipment/system
- ◆ Purchase of reliable equipment with better maintainability aspects
- ◆ Use of skilled and trained operators with proper attention and attitude for the work
- ◆ Use of correct type of spare parts and lubricants
- ◆ Timely availability of equipment to maintenance staff
- ◆ Timely availability of special facilities required
- ◆ Quality control of maintenance functions

Due attention to the above points will increase the efficiency of maintenance functions at all levels and will also provide satisfactory work.

UPDATING MAINTENANCE ORGANIZATIONS

Due to rapid advancements in technology, changes in the maintenance set-up have become essential. There should be flexibility in the maintenance organization to accommodate such changes. Special types of equipment are always required to speed up the maintenance work and therefore the maintenance engineer must keep in mind their utility in view of the changing technological environment. When a new organizational set-up is formed, the new equipment/facilities should be planned to be installed at the place, where most of the maintenance work is concentrated. This will reduce the equipment movement for maintenance work. For example, if a new testing machine is to be installed due to new requirements it must be placed, where other similar types of machines are installed in the workshop to conduct maintenance work on the related group of machinery.

With the addition of new monitoring techniques it is possible to assess the successful life of a component/equipment. Hence it is desirable to make use of all such facilities as far as possible. This can ultimately improve the performance of the equipment with a reduction in total maintenance costs. The quality of maintenance work will largely depend upon the skill of the working personnel as well as the quality of the spares used.

The modern maintenance organization should have all the advanced testing and monitoring facilities to expedite and achieve better quality of work. Extensive use of computers must be made to preserve the information related to maintenance work which can be used in future in similar type of equipment/system. Efforts should be directed to promote research in the area of maintenance where spare parts are not available particularly for the imported equipment/ system.

QUESTIONS

1. Outline the organizational set-up of the maintenance department in any industry which you may have seen and suggest changes that you would wish to implement.

2. Discuss the factors that influence the performance of a maintenance department.

3. Explain with examples how organizational problems can adversely affect the productivity of a particular industrial set-up.

4. Illustrate the main objectives of maintenance organizations that can help it in meeting the challenges of the maintenance activities.

5. What are the organizational requirements of a maintenance function? Explain briefly.

6. Discuss the merits and demerits of different types of maintenance organizations.

7. Describe the methods employed for cost minimization in any maintenance organization.

8. How can a maintenance organization be made more effective in today's industrial growth environment? Explain briefly.

9. Explain the measures required to update the existing maintenance organization.

Chapter

12

Maintenance Equipment and Facilities

INTRODUCTION

For successful completion of any activity, depending upon external support, it is essential that the same be implemented at proper place and time. The maintenance function depends upon many external factors and the ready availability of its services plays an important role in effective completion of the jobs. The major maintenance facilities include well-equipped engineering workshops and the stores organization. Depending upon the type of maintenance organization adopted, these facilities are provided both at the unit and corporate levels. Generally, small workshops and stores are provided at each unit to look after the daily requirements of the maintenance function. However, there are various advantages of centralized facilities which include the following:

- Cost effective utilization of manpower and equipment/ facilities available in the organization
- Better spare parts management through computerized inventory control
- Quality work with the effective utilization of modern maintenance facilities
- Standardization of work which is repetitive in nature to reduce total time of maintenance in future
- Reduction in maintenance/repair time, through effective planning, scheduling and quality control
- Scope for better research and development activities to analyze the causes of equipment failures and their remedial actions.

On the other hand, the greatest requirement of the centralized system is that it should be an efficient and effective organization with proper coordination and control. It should include well-organized and fully-equipped workshops.

Requirement of Modern Maintenance Department

High level maintenance requirements for the modern industry calls for the following facilities to be established in the maintenance department:

- ◆ Non-Destructive Testing (NDT) facilities for the surveillance of the plants and machinery
- ◆ Data bank on failure statistics and spare parts in the computerized format with some alternative solutions on maintenance problems
- ◆ Anti-corrosive treatment facilities to protect equipment/ system from environmental hazards
- ◆ Welding and crack detection techniques such as X-rays, Dye Penetrants, Radiography, Ultrasonic, Magnetic particle testing, Scanning electron microscope, Interferrometric holography, Eddy current testing and so on
- ◆ Rotating machinery such as centrifugal compressors, turbines, apparatus for dynamic balancing, vibration monitoring and high quality bearing setting services.
- ◆ Calibration facilities
- ◆ Supporting services

The other requirements are as follows:

(i) Technical literatures like maintenance schedules, maintenance instructions, operating manual, fault analysis charts/diagrams drawings and specifications etc.

(ii) Well organized spare parts system

(iii) Special tools and facilities/equipment etc.

The maintenance department should be responsible for the following:

- ◆ The maintenance policies of each group should be inline with the philosophies of the corporate objectives
- ◆ The existing maintenance practices must be reviewed in time to incorporate the latest techniques/methodologies
- ◆ Analysis of failures to know its impact on production and operating costs
- ◆ Development of detailed specifications for key equipment for their procurement by the purchase department
- ◆ All the working personnel must understand the concept of reliability as a continuous process for improvement

- Providing the craftsman with the procedures, technology and training required for verifying the quality of their workmanship in order to reduce the need for rework
- Utilization of maintenance information system for taking effective and timely decisions.

The role of well-equipped and coordinated workshops is important for successful completion of maintenance work and therefore their position/location in the plant will also be crucial and dominate factor.

WORKSHOPS

Since the workshops are the essential parts of the maintenance organization, their structures as well as the equipment located therein, largely depend upon the type of machines to be maintained. However, the general layout of the workshops should be more or less of similar type. The basic characteristics of a good workshop are the following:

- It must have all the basic facilities to enable effective utilization of equipment used in the plants.
- It must have all the facilities for carrying out minor modifications to parts of the equipment.
- It must have provisions to ensure quality of raw materials to be used for maintenance work.
- It must have a research and development wing for devising new methods of testing and repair of the equipment/system.
- It must have the requisite facilities for training the maintenance personnel working in the field to update their knowledge.
- It must maintain a proper record system for the utilization of manpower and equipment at its disposal.
- It must have sufficient place for future expansion to accommodate modern facilities.

Sometimes it is not possible to have all the above facilities in the workshop for maintaining the equipment. For the major tasks of overhaul and preventive maintenance, the facilities not set up due to financial constraints can be subcontracted to original equipment manufacturers or their specialized persons called in at the site to help in major repairs. Manufacturers of capital-intensive equipment also provide workshops at important places to facilitate the maintenance activities of the customers.

Workshop Classification

Depending upon the nature of maintenance requirements and the organizational set-up, workshops can be divided into various groups such as:

- ◆ Central workshop
- ◆ Unit workshop
- ◆ Apprentice workshop
- ◆ Auxiliary workshop
- ◆ Mobile workshop/trains
- ◆ Commercial workshop
- ◆ Special workshop
- ◆ Field workshop

The central workshop is normally well equipped with modern test and repair facilities. It usually has a big machine shop with different types of lathe machines, milling machines, drilling machines, grinders, and other ancillary machines. In the central workshop, specialized types of jobs are taken for manufacturing or modification. The central workshop also has the facilities like welding, painting and even heat treatment. Small research and development groups are also attached to such workshops to carry out the development work, particularly in the area of manufacture of imported parts, which are difficult to get from the original manufacturers.

The unit workshops are attached to smaller establishments and have basic facilities of machining, welding, etc. where only limited maintenance works can be undertaken.

The apprentice workshops are equipped with facilities on a moderate scale for the purpose of training the technicians. All the basic facilities of the machine shop, welding shop, carpentry shop, smithy shop, electrical and instrumentation shop, battery shop, and electronic shop, etc. are made available in these workshops.

Auxiliary workshops are workshops in the conventional sense that have some facilities under temporary sheds to cater to the maintenance requirements at the site where the equipment is actually used. This is only for saving the time to rectify breakdowns. Trained personnel are placed at these places to deal with such eventualities. They can also be used to carry out those weekly or monthly planned maintenance programmes for which shop facilities are not required.

Mobile repair workshops are useful for undertaking running repairs and minor breakdowns at the site where, sometimes, it is not possible to shift the equipment to the workshop for the maintenance work. For example, if the tyres of a vehicle need replacement due to some reason, then they have to be replaced at the site or it may be that a broken axle shaft has to be attended at the site itself. Mobile repair workshops are used in the case of railway breakdowns, or automobile failures on the road, or in military operations at the forward landing areas.

The commercial types of workshops are very common and are found almost everywhere. These workshops undertake different types of repair work, which include repair of gears, pistons and cylinders of automobiles. In these shops, the damaged parts of equipment are also repaired, besides providing contract service to the users. The organizations, which do not have their own workshops, mostly depend on these commercial shops for several types of repair work.

Special workshops are established exclusively for a particular type of repair work. For example, in an automobile workshop the work related to a particular type of automobile may only be carried out. Similarly an electric shop may undertake the repair of only electric motors. These shops aim to produce quality work since trained persons are normally employed in these shops. They are becoming very popular because of their quality work and timely service.

The field workshops are established for equipment, which are bulky and heavy in nature and cannot be moved easily from the working area. For such equipment minor nature of works are attended by these workshops to enable their movement from working area to the central workshop.

A simple workshop organization can be of the following type (Figure 12.1).

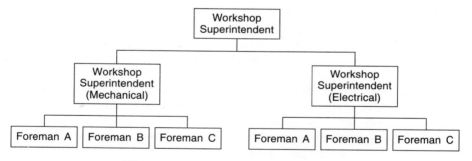

Figure 12.1 Workshop organization.

Workshop Location

The workshop location is one of the important parameters for the maintenance function particularly when the frequency of failures is high and the equipment cannot be shifted very frequently. The selection of a workshop site should be based on the following criteria.

◆ Workshop should be located centrally in a plant so that all the concerned departments do not have much difficulty in transporting their equipment

◆ Adequate provision should be made for future expansion and modification of the workshop to accommodate new and advanced facilities

- The workshop should be easily accessible by road and rail so that even large-size equipment can be handled and transported without difficulties
- The workshop site must have the essential facilities such as electricity, water, gas, and drainage
- The movement of men and material should be designed to be minimum during the course of work in progress
- Arrangements for proper light and ventilation must be provided
- Exhaust fans must be provided for circulation of fresh air and removal of intoxicated fumes and gases
- Provisions must be made to fight out fires in case of their occurrence, i.e. Fire extinguishers must be placed at proper positions
- Arrangements must be made to protect high-tension electric cables used for running the machine in the workshop

Central Workshop Organization

The central workshop is the most important workshop for any organization. This workshop consists of a number of maintenance shops within it. The extent of facilities incorporated in these shops depends on the maintenance requirements. Mainly the type and size of the equipment to be maintained determine the size of this workshop. Normally the following shops are essentially present in any central workshop:

Machine shop: It is one of the most important shops of a workshop, which is used not only for repair work but also for the manufacture of parts and other fabrication work. Lathes, milling machine, drilling machine, grinders, etc. are housed in this shop. Other facilities such as tool room, smithy, welding, forging and carpentry are also provided in a machine shop.

Electrical repair shop: Repair and maintenance of electrical equipment can be carried out here. The testing facilities for all types of electrical equipment must also be provided in such shops.

Engine overhaul shop: Such shops carry out overhauling/ repairing of engines. Specialized persons are normally employed in such shops. An industry deploying several vehicles usually maintains a separate engine shop. This shop should have all testing and repair facilities required for the engines.

Test bench: After the overhaul of engine it is required to be tested for its performance and therefore arrangements must be made to establish test bench. Here the engine can be tested with the application of desired loads with actual fitting on the machine.

Hydraulic and pneumatic repair shop: These shops have several types of repair/overhauls facilities for hydraulic and pneumatic equipment used in an organization. Facilities like hydraulic trainer and pneumatic test rig are made available in these shops to test all such equipment.

Welding facilities: Almost all mechanical and electrical equipment have some use of welding and therefore provisions must be made to provide this facility during maintenance work as well.

Additional facilities: In an advanced central workshop, numerically controlled machines are also made available for better finish and accuracy of fabricated parts. Heat treatment facilities are also provided in some of the workshops depending upon their requirements. No shop is complete without good quality tools and special test equipment, which may be required for the maintenance function. There are a few advantages of a centralized layout and they include:

◆ Better supervision of work
◆ Good communication between different shops
◆ Transportation time from one shop to another being reduced to minimum
◆ Better planning and sequencing and control of various operations in shops
◆ Better utilization of available space and facilities

The biggest drawback of the centralized workshop layout is the interference between its own shops, noise and general pollution, which may influence the working efficiency of workers. The other disadvantage can be the formation of groups, which may lead to conflicts.

Layout of Workshops

Before laying out any type of workshop provisions must exist for proper layout of machines and those of future expansion. The layout also depends on the type of shop, which in turn depends on the type of equipment to be maintained. The shops like plating, painting and annealing are housed away from the main shop. This is for the reason that such shops emit toxic fumes/gases which are harmful for the working personnel in the workshop. The total area required must be calculated taking into account the present maintenance load requirements of the organization.

The different shops must be laid in such a manner that the work progresses in a forward direction as far as possible through the shortest possible route from the point of entry to the point of its completion. This will prevent unnecessary handling and transportation of material and spare parts required for the maintenance work. Figure 12.2 shows the typical layout of an engine overhaul shop. Provision for spare parts can also be made at a convenient place.

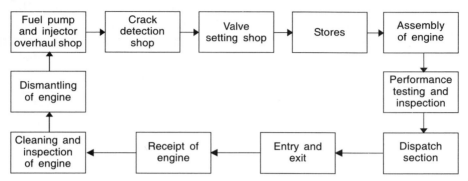

Figure 12.2 Typical layout of an engine overhaul shop.

It can be seen from the above layout that the movement of the jobs does not affect the working of the other departments/sections. Here it is easy to control the repair/maintenance work. Wherever extra manpower is required, it should be provided to complete the job so that the sequence of operations can be maintained and the work does not get delayed.

In the receipt section the job is taken over with all its accessories and requisite documentation and then passed on to the inspection shop, where the nature of the work content is investigated. The engine is then dismantled and its subsystems/parts such as fuel pump, injector etc. are sent for overhauling to different bays/shops. The engine block is thoroughly checked for any cracks in the crack detection shop. The valves are sent to the valve setting shop. Any parts which need replacement can be demanded from the stores. Dimensional accuracy of the components is also verified for proper working of the engine. Finally, the reassembly of the engine is undertaken in the assembly shop, after which the complete engine is tested for its performance in the testing shop. All through the quality inspectors also check the work. On completion of all work and quality checks, the tested engine is then dispatched to the concerned department/section.

Similar types of layouts can be planned for different types of components/parts such as electric motors, hydraulic motors, pumps, and so forth in different overhaul shops.

Workshop Control

Normally, the workshop superintendent, who is the overall in-charge, looks after the administrative control of a central workshop. It is the duty of the workshop superintendent to assign the maintenance work to the right person at the right time. His coordinated effort has to be geared towards improving the availability of the production/service equipment. He is directly placed under the chief engineer. The financial control of the workshop is implemented through its annual operational budget prepared by the workshop superintendent in consultation with his immediate subordinates and the finance manager.

To effectively utilize the resources of the workshop, it is essential to plan and coordinate the maintenance work in advance. A senior foreman should carry out supervision of work in each shop. From time to time the workers should undergo training in the related areas of maintenance and control and learn new techniques of inspection/monitoring such as non-destructive testing (NDT), and so forth.

STORES

Stores form an important part of the maintenance function. For completion of maintenance work on time, the supply of the spare parts must be maintained regularly through the stores organization. Before setting up a store, classification of the equipment must be done to identify each machine and its particulars and the particulars of its spare parts. The equipment history cards also help in planning the stores. The working conditions of equipment may also affect the requirements of the spare parts. For efficient working, it is advisable to store the items category-wise, for example, mechanical, electrical, hydraulic, and so forth. The duties of the stores in-charge are very crucial since success of the maintenance organization largely depends on ready availability of spare parts.

Stores Classification

Depending upon the work requirements, the stores may be classified into various categories such as:

- Central store
- Workshop store
- Section store
- Special store
- Field stores

A central store is normally established to meet the overall requirement of the spare parts, where items of all the categories are stocked. It is the responsibility of the central store to procure the items as and when projected by the users. Some buffer stock of critical items must be maintained to deal with the emergency demands. Proper methodology has to be adopted for timely procurement of the items in sufficient quantities. The quality of the items must also be given due consideration.

Workshop stores form an integral part of the central workshop and maintain stock of spare parts and materials required by the central workshop as per the production plan.

Section stores are provided to cater to the day-to-day needs of the shops and therefore only small quantities of spare parts are stored in these shops. They form the auxiliary stores of a workshop to minimize the routine workload of the latter.

Special stores include items such as fuel, oil, lubricants, and the like which need to be stored in a safe place away from the normal place of work.

The necessity of field stores arises where heavy equipment used in the field cannot be transported easily in to the workshops or bulky items which have difficulty in their shipment.

Location of Stores

The location of central stores, depending on the availability of place, should be accessible by road/rail transport facilities and be in close vicinity of the central workshop as far as possible. Types of store items also decide the location. For example, the inflammable and explosive items have to be stored away from the domestic areas. At the same time, bulky items such as construction materials need to be stored in open place. Normally, a store should be a single-storied building, though electrical and electronic items can also be stored in multi-storied buildings.

Store Layout Design

The size of any store will depend upon the type of equipment and their numbers. As spare parts storing is expensive, the quantity of the parts must be decided carefully based on statistical analysis and the past experience on inventory requirement of similar type of equipment. The storage area should be well planned to accommodate bulky as well as small items properly, and space should be earmarked for future expansion.

In designing a store layout, due consideration should be given to storing sensitive items to protect them from damage caused by moisture, light, and fire. The danger of theft of small and valuable items should also be taken into account. Such items should be stored in a safe place. Special care must be exercised towards storing the perishable items so that they can be utilized well before their expiry date. Congestion of stores should be avoided to obviate the chances of damage. Similarly if the items are to be stacked, care should be taken to avoid damage to the parts.

In case the store building has multi-storey, the light items must be placed in the second storey and the heavy items in the ground floor. There should be proper arrangements for the movement of the stored items, i.e. stairways and lifts must be provided. For transporting the spare parts, forks, light trucks should be made use of. To store small parts of machines, the bins and shelves must be provided. The bin size must match the size of items to be stored in it. Bins must have the reference number of the parts for easy recognition of the items. This will help the users to identify a particular item/component quickly.

QUESTIONS

1. What are the main considerations taken into account in the design layout of a workshop?

2. How would you decide the location of a central workshop for automobiles being used at different locations?

3. What is the role of the stores organization in achieving effective maintenance? Explain.

4. Why are maintenance-related facilities required, when they all cause extra burden on the organization? Discuss.

5. Describe the classification of workshops, giving merits and demerits of each type.

6. Explain the importance of workshop location for a maintenance organization.

7. Discuss the importance of a good workshop layout for a maintenance organization.

8. On what basis are the spare parts stores classified? Explain.

Chapter
13
Maintainability

INTRODUCTION

Maintainability is a concept, closely related to the characteristic of equipment design and installation. It is expressed as the probability that an item will remain in a serviceable condition for a given period of time or restored to a specified condition within a given period of time. Maintainability is also expressed in terms of the minimum cost of maintenance as well as the accuracy of the maintenance functions.

The basic objective of maintainability is to design the equipment that can be maintained easily in minimum time and at minimum cost. This implies that the requirement of other supporting resources such as spare parts, manpower, and facilities of tools and test equipment must also be minimal.

The concept of maintainability was developed to provide the designers with special knowledge related to the support and maintenance of equipment. The need of this was felt due to growing complexity, size and rapidly changing technology in respect of design and fabrication of equipment and today maintainability has become one of the important goals of design.

The objective of maintainability requires the involvement of the complete design procedure including planning, production, installation, and actual working of the equipment. During the planning stage, the maintainability requirements are defined and incorporated in the design. The next step is to consider the functional aspects of the designed equipment; these designed characteristics are then measured or evaluated to assess the areas of improvements. In this way the particular support requirements of the equipment are controlled to meet the operational needs.

The realization of the objective of maintainability requires involvement in the total design process. This includes participation in advance planning, design development, production, and activation of equipment or system. Applied maintainability consists of four discrete work stages within each phase of the design process. These are (1) planning, (2) design application, (3) measurement, and (4) evaluation. In planning stage, maintainability requirements are defined and translated into design criteria. Next, the criteria are applied to the design to establish the desired inherent functional and physical characteristics of the equipment or system. These characteristics are then measured to verify the goals and objectives and finally, evaluation is made for any improvement.

Equipment that is easily maintainable will have better market than the one, which is difficult to maintain. Therefore maintainability plays an important role in marketing a particular equipment. In the developing countries like India, this issue is more relevant because the concept of obsolescence is not strictly followed and equipment is continued to be in use for as long a time period as possible.

The modern practice of maintainability originated during the World War II from the numerous studies launched by the United States defense services and was based on the demand for highly reliable battle-worthy equipment for complete success of a mission. Though the reliability programmes were aimed at prolonging the operational availability of equipment, the maintenance requirements of these equipment could not be wholly eliminated. On the other hand, with changing technology and the smaller size of equipment, the maintenance cost also increased in proportion to the operation cost. With this in mind, the idea of maintainability was conceived to reduce the product's life cycle cost of maintenance.

During the early stages of the maintainability concept, maintainability was associated with cost effectiveness, system effectiveness and provision of supporting services for efficient performance of the maintenance function. But at a later stage, the concept of maintainability was built into the design process, system planning and systems analysis because of its increasing importance and need.

The primary considerations governing any maintainability plan are as follows:

♦ The designer's role is the most important as many alternatives are possible during this phase and the one with optimum results should be considered

♦ During the design phase, maintainability requirements for a particular equipment can be fully specified

♦ Maintainability features can be carefully incorporated into the product during the design and fabrication stages

♦ At the time of the maintainability implementation programme, reliability and other characteristics of a system can be evaluated

♦ Good maintainability provisions can help the maintenance department to carry out maintenance successfully with proper cooperation of the users of the equipment.

◆ Good maintainability can also improve the safety of personnel.

Any equipment designed and fabricated should be cost effective. Therefore, it is important to note the benefits achieved by implementation of maintainability. The acquisition cost includes the money spent to purchase the item for specified performance. The operational cost is the money spent to keep the equipment in operational use for a planned period of life. This also includes maintenance cost, which can be minimized by proper implementation of the maintainability concepts especially in complex and costly equipment. For example, whenever frequent changes of a component are required, it should be easily accessible.

The system effectiveness is defined as the probability of system success. Its objectives are: (a) system readiness and (b) achievements of desired results. The system effectiveness can be classified into three areas:

1. *Availability* (operational) determines the probability that a system is ready for use in terms of mean time between maintenance (MTBM), mean downtime (MDT), and ready time. The MTBM is a function of reliability, the MDT is a measure of maintainability, and the ready time is also a function of reliability. The interaction of these parameters determines the system availability.

2. *Dependability* determines the probability of successful performance of a system/equipment during its planned period. This will vary from one type of system to another, for example, the measurement of performance of a surface-to-surface missile will differ from a manned aircraft vehicle or a remotely operated vehicle.

3. *Capability* of a system depends on its reliability. Sometimes an efficient operator can obtain better results from the same type of equipment, for example, the racing cars or even motorcycles.

MAINTAINABILITY AND ITS COST

Incorporation of higher maintainability increases the cost of production of any machinery. However, it reduces the operating cost considerably. As already emphasized, maintainability of any machinery needs to be considered at the design stage. A number of factors influence the degree of maintainability in a machine. These include:

◆ The need for completion of work in minimum time with minimum number of maintenance personnel.

- The skills needed of the working personnel should not be too high.
- The consumption of spare parts should be minimal.
- The requirement of tools and test equipment should be as less as possible.
- Ease of working.
- Less number of outsourced services needed.

The above parameters, if considered during the design phase, can improve the cost effectiveness of the system. Depending upon the design requirements, the support systems can be planned such as provisioning of spare parts and the like. Though maintainability is invariably associated with a higher cost of production of a machine, it nevertheless promotes sales because of its overall cost effectiveness.

The function of reliability and the working conditions required are both closely related to maintainability. They have a significant impact on implementation of maintainability. Whereas a reliability programme helps to extend the useful life of equipment, the human interface between the physical and functional design features also supports the system. The human engineering factors directly influence the mean time between maintenance and the mean down-time of the equipment. They also help in identifying the manpower requirements, depending upon their level of skill and training, as well as contribute to economic aspects of the machine.

Maintainability contributes to the minimum outlay of funds for system acquisition and utilization through attainment of objectives associated with maintainability design and support planning.

The additional factors to be considered in the placement of a maintability organization are as follows:

(a) The maintainability organization must have status at par with other sections such as logistics, technical publications, design engineering, systems engineering, training etc.

(b) The maintainability organization must be identified and their functions should be known to all concerned.

(c) The maintainability organization must have authority as well as responsibility at par with other departments.

(d) Maintainability should have organizational ties with reliability, safety, human factors, value engineering, and test group functions.

(e) The location of the maintainability organization should be in close proximity of logistics, technical publications, system engineering, design engineering, training etc. of the parent organization.

MAINTAINABILITY ANALYSIS

Maintainability analysis correlates qualitatively the process of maintenance requirements with the specific equipment design requirements in terms of quality. Such analysis can provide:

- ◆ Specification of maintainability requirements to be considered during the design phase of equipment.
- ◆ Details of checklists to ensure that the appropriate features, as required, have been incorporated.
- ◆ A summary of basic maintenance functions and support requirements for the equipment/system.
- ◆ Specification of supporting services needed to operate and maintain the equipment.

Maintainability analysis is carried out before the actual design of the equipment, i.e. during the system development cycle.

FUNCTIONAL ANALYSIS

A functional approach must be applied as a frame of reference for identification of initial requirements at each design stage of the system. This defines the basic operational and maintenance functions and also ensures that:

- ◆ All elements of the system have been identified and defined.
- ◆ All means of support requirements for the maintenance function have been identified and defined.
- ◆ A common reference point for all the design and development activities has been formulated.

In the system engineering approach, the first step is the functional description of the system to understand the basic needs of all the components/parts. It represents the level of all functions to satisfy the total system requirements. Once these functions are defined, they can be translated into function flow diagrams. All operational functions of a system are specified and then divided into sub-functions for simplicity.

Once the operational functions are known, the system engineering process proceeds with the development of the maintenance functions. With every operational function, some performance requirements are associated. A check of a particular function will indicate the applicability of the next function and also provide detailed maintenance functions and their problems associated with the system. These maintenance functions will reflect the considerations of the customers concerning the downtime of equipment and maintenance resource requirements. Thus, this kind of system engineering process continues till the basic design configuration is established with all the alternatives possible at the design level.

The functional analysis must take care of the following:

◆ Identify the requirements for preventive and corrective maintenance

◆ Identify the requirements for equipment handling during its actual working

◆ Identify the maintenance tasks and describe methods, facilities and equipment required to carry out various maintenance activities

◆ Identify special instruments, safety and support requirements for the maintenance work

◆ Project the criticality of the maintenance effort for system/equipment readiness for effective performance

◆ Identify the maintenance task frequency to reduce the downtime of the equipment/system

◆ List the spare parts and manpower required for each maintenance support task.

Maintainability allocation: When maintainability programme plan has been prepared and functional analysis and maintenance data have been updated, design and performance specifications for the system are prepared. These specifications become the primary means whereby maintainability requirements are incorporated in every specification. To determine quantitative maintainability requirements for each item, engineer concerned must budget the MTTR of the various equipment items such as that the statistical mean is equal to or less than the overall MTTR requirement. This budgeting is accomplished by means of maintainability allocation.

All allocation consists of determining the contribution of active downtime of each equipment item, and evaluation of the contributions for all such equipment against the established MTTR for the system. In the event of the MTTR allocation for the system exceeds the system's MTTR goal, the maintainability engineer must revise the allocation.

MAINTAINABILITY PREDICTION

The prediction of the expected number of hours that a system or device will be in an inoperative or 'down state' while it is undergoing maintenance is of vital importance to the user. Therefore, once the operational requirements of a system are fixed it is desired that a technique be utilized to predict its maintainability in quantitative terms as early as possible during the design phase. The prediction should be updated continuously as the design progresses to assure a high probability of compliance with specified requirements. The advantage of maintainability prediction is to point out the areas of poor maintainability, which justify the product improvement, modification, or change of design. Another useful feature of maintainability prediction is that it permits the user to make an early assessment of whether the predicted down-time, the quality, quantity of personnel, tools and test equipment are adequate and consistent with the needs of the system.

A good prediction technique should be one in which the procedure itself can be used directly to pinpoint design deficient or support-deficient areas for improvement. It is important to decide the time of prediction, which can start at the time of contract award.

Elements of Prediction

The following are the elements of prediction:

1. Failure rates of components/items at the appropriate time to establish frequency of maintenance.
2. Repair time required at maintenance level for different types of maintenance systems.

Methods of Prediction

The prediction method is highly complex and requires a combination of a mathematician/maintainability engineer and therefore should be on the requirements of the same. This will also depend upon, applicability, point of application, basic parameters of measure, information required and constraints if any.

DESIGN FOR MAINTAINABILITY

In the present context it is desired that equipment must be reliable and dependable requiring minimum maintenance and consequently suffering from least downtime even if the initial cost may be high. For achieving the optimum maintenance results, special care should be taken in the design of equipment. A proper design can eliminate many maintenance-related problems. Selection of the correct materials is one of the important factors. The design of the modern plants should be such that the operating periods be limited rather than the availability of the equipment. The specified standards must be followed in terms of safety wherever possible, which may also help in cutting down the maintenance requirements.

Equipment must be laid out in such a way that there is sufficient space for carrying out repair and maintenance work. Accessibility is an important factor as far as maintenance work is concerned. Space must be available for removing and replacing the components. Sometimes it is seen in practice that extra time is spent in locating the faulty parts/components, because of their awkward locations. If a particular part is likely to fail frequently, it should be easily accessible, for example, a fuse in an electrical circuit.

Maintainability is applied to four discrete work stages within each phase of the design process:

◆ Planning
◆ Design application
◆ Measurements
◆ Evaluation

In the planning stage, the maintainability requirements are defined and incorporated into design criteria. The criteria developed are applied to the design to yield the desired inherent functional and physical characteristics of the equipment/system. These design characteristics are measured to verify the quantitative and qualitative goals. And finally, the design and the results of measurements are evaluated to assess the areas of improvement.

Fabrication of Equipment

After completion of the design, it is important to fabricate the equipment/system so as to check whether it meets the operational requirements. At this stage if some problems are noticed in connection with the maintainability of the equipment, the same can be referred back to the designer. With minor modifications, the major problems of the maintenance function can usually be minimized. Sometimes, even the material problems can be sorted out at this stage.

COST EFFECTIVENESS

It is the measure of the equipment system value compared to that of the other similar types of equipment, where the maintainability criteria have not been incorporated successfully. Such measures reflect the cost of purchase, installation and maintenance of the system over the period of its useful life, and the capability of the equipment to perform its intended functions. Cost effectiveness (CE) can thus be expressed as:

$$CE = \frac{\text{System Effectiveness (SE)}}{\text{Acquisition Cost} + \text{Utilization Cost}}$$

The system effectiveness is defined as: "The probability that the system can successfully meet an operational demand within a given time period when operated under the specified conditions."

The acquisition cost of the equipment includes the money spent to attain the desired performance, besides the installation cost and miscellaneous expenditure incurred towards training and support equipment.

The utilization cost of the equipment/system refers to those funds, which were spent to maintain the equipment in the operational condition during its planned life. This includes the maintenance cost and the general operating cost. Therefore, cost effectiveness can be visualized as the part of maintainability.

Maintainability Training

The objectives of the maintainability-training programme are (i) to train engineers in the performance of tasks and (ii) to acquaint other company personnel with the purpose and functions of maintainability organization. The following type of courses can be undertaken:

♦   *Familiarization course:* It should be conducted for all programme management personnel and supervisors who are involved with maintainability programme and duration should not exceed from one hour at a time.

♦   *Design orientation course:* Here all design engineers should be briefed on maintainability criteria documents, mainte- nance policies, maintenance design criteria and programme design standards.

♦   *Course for support personnel:* All the associated personnel must be briefed about the maintainability organization and its objectives on the interpretation and use of maintenance analysis. The emphasis should be placed on what data are contained in the analysis report, purpose of each data entry and how they are applied to related products.

QUESTIONS

1.   Discuss the need and importance of maintainability.
2.   How does the cost parameter play its role towards achieving the desired degree of maintainability? Explain.
3.   Why is maintainability analysis required? Briefly discuss its need.
4.   What is the importance of design in achieving the desired level of maintainability? Elaborate.
5.   Explain the needs of training in area of maintainability briefly.
6.   Describe the relationship of the maintainability analysis function and the maintainability system design function.

Chapter

14

Lubricants and Maintenance

INTRODUCTION

The main function of most lubricants used is to minimize friction and wear between two mating surfaces and to extract heat. They also have to remove debris from the contact area, e.g. combustion products in case of internal combustion engines, swarf in the case of metal cutting operations. Sometimes, they have to protect the lubricated or adjacent parts from corrosion, but it is not the prime function of the lubricants. On the other hand, many lubricants contain corrosion inhibitors.

The failures of mechanical equipment system increase in number when the same gets older in use as well as with the passage of time. And therefore friction and wear are the most common problems associated with the industrial equipment which can be minimized with the use of proper quality of lubricants. Wear is a progressive deterioration of a surface with loss of shape and weight and is associated with reduction in operating efficiency, increase in power losses and eventual breakdown of machines demanding replacement or major repair. A wide range of materials with varying characteristics is available to protect against wear. The reduction in wear is possible through a timely and proper lubrication system. This chapter gives a brief description of lubricants related to the maintenance function.

Lubricants are used in almost all mechanical systems and equipment and they play a very important role in maintenance engineering. The proper use of lubricants contributes towards increased life of equipment and plant, and leads to reduction in downtime and maintenance cost. At the same time, the lubricants assist in ensuring the safety of operations. For example, the shortage of oil/lubricants in either brake system or gear system may cause serious problems in the operation of the vehicle. It has been observed that lack of proper lubrication leads to disturbing sounds emanating from the moving parts of machines. This often irritates the operator of the machine. To achieve a sound quality of maintenance, therefore, the maintenance management needs to take care of the following:

◆ To ensure that all the components/parts are provided with proper type, quality and quantity of the lubricants during the maintenance of the system.

◆ To select the correct type of lubricants as well as the lubrication systems to improve the operating conditions of the equipment/machinery.

◆ Periodic monitoring of the quality of the lubricants, which are in use and their replacement within the specified time period.

◆ Proper storage and handling of lubricants to maintain their quality and sustain their useful life.

LUBRICATION SYSTEMS

Lubrication systems of equipment mainly depend on the design as well as the operating conditions of the equipment. In the areas such as cold or hot, the lubrication system has to be different for the given two conditions. One may require lubricant that can withstand high temperatures and the other one may need a lubricant, which does not freeze at low temperatures. They also depend on the method of applying the lubricants. Though differing in their working principles, the purpose of the lubricants is the same and in general classification, the lubrication systems can be divided into the following groups.

◆ Manual lubrication systems

◆ Automatic or self-designed lubrication systems

◆ Built-in lubrication systems

The lubrication systems can also be classified as:

(i) Open lubrication system, and

(ii) Closed lubrication system

In the case of manual lubrication, hand grease guns and oilcans are used for the application of the lubricants. The oil or grease must be filled in the cans in a non-dusty atmosphere to avoid contamination. Where grease is used as the lubricant, mostly one type of nipples should be used as far as practicable, since it will ease the greasing operation. The oilcan should also be dust proof and fitted with a cover. The limitation of manual lubrication is that the chances of contamination are more, sometimes even due to negligence of the worker, since the part of the lubricant is exposed to the atmosphere.

In modern equipment, automatic lubrication systems are provided wherever feasible. Such a system protects the lubricant from adverse atmospheric conditions. There are different types of systems being used such as, waste pad, sight feed oilers, wicked oilers, gravity systems, mechanical force feed oilers, lubricators, hydrostatic lubricators, splash oilers, etc. The more recent systems can be classified into the following:

- ◆ Centralized lubrication systems (manual)
- ◆ Centralized lubrication systems (power operated)

In a manually operated lubrication system, the lubricator discharges the lubricant through a main tube to feed the branch tubes connected at the tube points. Whereas, in the case of a power-operated lubrication system the lubricant is pumped by its own pump to several lubricating points on a time controlled cycle. This system has been very successful and is employed today in many heavy-duty machines that are used in mines. The main advantages of this system are as follows:

- ◆ Positive lubrication
- ◆ Reduction in the idle time of the machine
- ◆ Reduction in the number of workers employed for lubrication tasks
- ◆ Lower maintenance cost

In the power-operated lubrication system, the chances of failure of equipment are reduced, which are more in the case of the manual lubrication system due to neglect of the operator, that is, not providing lubrication in time and also the sufficient quantity. But at the same time, failure of the power-operated lubrication system can be very serious, since it may cause seizure of the operating machine/system. However, some safety devices are being provided with such systems.

In some of the cases, it is practically not possible to provide lubricants to the moving parts. Sometimes the conventional methods of lubrication cannot be used. In such cases the sliding surfaces are made of materials which themselves provide some lubrication in the form of process fluid. Sometimes the moving parts have adequate wear resistance, which can be achieved by applying specially treated coatings. The bearings and seals, operated at very high temperatures, high pressure, and high speeds, fall under this category. The coatings and surface treatments are applied on the top foil and general surface to minimize wear of the rubbing surfaces. The selection of suitable coatings depends upon the working environment of the equipment.

In the case of open lubrication system, the same is applied through external agencies such as grease guns and oil canes, whereas in the event of closed lubrication system, it is applied within the system itself. For this purpose, due attention is to be focused while the system is under the design process.

Selection of Lubricants

It is the normal practice on the part of the manufacturers to recommend the type of lubricants most suitable for their equipment. Such recommended lubricants must always be used. It is a good maintenance practice to prepare a chart of lubricants for important components recommended by the manufacturer and display it near the concerned machine. The proper use of the lubricants pays in the long run by way of increased life and less breakdowns of equipment/ parts. Therefore their selection is of immense importance. The following factors must be considered while selecting a lubricant.

- ◆ Type of the lubricant—solid or liquid
- ◆ Place of use, from the viewpoint of temperature considerations
- ◆ Environmental conditions—open to atmosphere or sealed
- ◆ Frequency of application
- ◆ Life of the lubricants

Planning of Lubrication Schedule

Planning of the effective lubrication system is very important since lack of proper attention towards lubrication of the parts/equipment often leads to a serious problem such as seizure of an engine or damage to the parts. To obviate such adverse effects, it is important that proper thought to lubrication be paid in the following manner:

- ◆ Prepare a master chart for lubrication of each equipment machinery, item-wise and also the due date for replacement/ refilling of lubricants. This must be displayed in the office of the foreman in-charge maintenance and all other important places where the workers can also see it.
- ◆ Workers must be properly assigned the task of lubrication to complete the work as per schedule.
- ◆ Record of every activity relating to the lubrication system must be maintained so that exigencies may be overcome when a particular operator does not report for duty. The information system for lubricants must be clear and sound to avoid such situations.

The record of lubrication can be maintained as follows:

(i) Records of recommended lubricants and their annual requirement.

(ii) Consumption of each lubricant equipment-wise.

(iii) Test records of lubricants.

(iv) Type of the lubricating system used in each equipment.

(v) The persons maintaining the lubrication system must be properly trained to understand the new and improved technology in the area of the said system and maintenance in general.

(vi) New developments in lubricants and lubricant systems must be followed in order to minimize the maintenance cost.

Handling Lubricants

In a large organization several different types of equipment are normally used which require a variety of lubricants. Contamination will affect the quality of the lubricant after it is produced. Due care should, therefore, be taken in its packaging and transportation. It should be packed in clean and airtight containers. The date of manufacture should also be mentioned on the container to avoid confusion during its use. Lubricants must be safeguarded from getting contaminated due to atmospheric effects, with passage of time. The basic objective should be to maintain their quality.

Handling of lubricants is very important as improper handling of containers can cause serious problems such as fire hazards. Leakage of lubricants from containers will not only cause loss, but also make the place dirty and slippery. The lubricants should not be exposed to high temperatures as that may deteriorate their useful properties and make them unsafe for use.

The conditions, which make lubricants unsuitable for use are: dirt, water, mixing up with other lubricants, high temperatures as well as low temperatures. Some of the lubricants have specified shelf life and their storage for longer periods may thus spoil them. Therefore, it is necessary to adopt an efficient storing system for lubricants and also use them in time.

During transportation of lubricants, due care should be taken so that the containers do not get damaged. For this, the single-tier storage system must normally be followed. Unloading of the drums or containers should also be handled properly to avoid any damage. Sometimes wooden platforms are used to avoid damage. The same practice should be followed for handling the lubricants in warehouses as well.

Storing Lubricants

It should be a normal practice to keep the lubricants in a shaded area so that they are not exposed to direct sunlight or rain. Labels must be put on all the containers or tins, so that due care can be taken at the time of issuing the lubricants. Storing the lubricants outside in the open may result in their leakage and loss due to varying temperature conditions. This may cause seals of containers to become loose and thus increase the possibility of contamination due to rain and dust, etc. The cold weather may also adversely affect some soluble oils and compound oils when stored in an open space.

As a temporary measure, the lubricants can be stored outside under a shaded area or under tarpaulin covers. The floor area where the lubricants are stored must be well maintained and the drums must be placed on wooden platforms to prevent their corrosion, which may spoil the lubricants over the course of time. Regular inspection of the store must be carried out to check for any damage or leakage of the containers; the identification marks must also be checked regularly.

Drums of different types of lubricants must be stored separately and suitable labels must be affixed to each drum to indicate the type of lubricant stored. Normally, the single-tier system is adopted for storing drums, otherwise wooden planks must be provided in between the two drums. While handling the drums, metallic rods or any other sharp objects which may cause damage to the containers, must be avoided.

In indoor storage of lubricants, due care should be exercised to maintain a fixed temperature inside the warehouse in all the seasons. Precautions should also be taken to avoid fire hazards. Provision must be made to implement the 'first in first out' concept for the stored lubricants in order to avoid contamination caused by environmental conditions.

Sealed drums can be stacked horizontally on wooden planks. Forklifts may be used for handling the drums/containers. The storeroom must be kept in the cleanest possible state. Its size will depend upon the consumption of the lubricants in a plant.

While using drums containing lubricants, the following precautions must be taken:

(i) The drums must be placed on a concrete slab and oil withdrawn directly into measured oil containers so that it could be issued to maintenance personnel.

(ii) The equipment used for lubricating purposes should be kept in oil racks so that their shape does not deteriorate when not in use; strict care must be exercised to keep the oil room and other dispensing equipment as clean as possible.

(iii) For draining oil from the drum, stop-cocks must be provided and kept in closed position when not in use. They should be checked for leakage from time to time.

(iv) Small trays must be kept underneath while draining oil from drums or to collect the oil drip, if any. Oil thus collected can be filtered and used.

(v) Grease drums/containers can be stored vertically and care should always be taken to keep them covered when the lubricant is not being withdrawn from them.

(vi) While draining lubricants from bigger containers, the normal practice should be to use measured containers for ensuring the correct amount of the product. The containers used for carrying lubricants from the store should be provided with lids/covers to avoid contamination of lubricants from dust and moisture during transit.

(vii) For drawing out grease from the drums, the use of hands should be avoided as far as possible in order to reduce the chances of contamination. Due care should be taken, not to mix up the different grades of grease by any means, which otherwise will affect the performance of the equipment.

(viii) For different types of lubricants, colour code schemes must be used to avoid confusion and trained persons must be employed for the job.

(ix) In the storeroom, the tags indicating the type and grade of the lubricant must be displayed for easy identification.

Coding Scheme

To avoid confusion while using lubricants, a colour code scheme should be followed for different types of grades of oils and greases. This will allow easy detection of a particular type of lubricant and also facilitate its storing. Various schemes of symbols can be used such as circle or triangle of different colours for different types of lubricants.

Maintenance of Lubricants

As stated earlier, to avoid contamination of lubricants it is essential to maintain the storeroom as clean as possible. To prevent penetration of moisture, the inside walls of the storeroom are painted. The floor of the room should be cemented so that it can be easily cleaned. No loose materials should be allowed in the room to avoid any contamination of lubricants.

The equipment used for handling lubricants must also be maintained with utmost care as far as cleanliness is concerned. Dirt will spoil the lubricants. As discussed earlier, it is also essential to maintain a coding scheme for the lubricants. Labels provided on the containers must be checked from time to time and changed if necessary to avoid confusion. For keeping the material and equipment, the places for each must be specified and this should be followed in practice for easy handling of lubricants.

The expiry dates of lubricants must be checked regularly to control wastage. Such perishable items must be stored separately with a clear indication of the same.

To avoid spilling, proper dispensing equipment must be used which may include pumps for oils, paddles and air-operated pumps for greases. Faucets are mostly used for drawing oils from drums of small size, whereas pumps can be used for large-size drums.

Though the items of handling equipment are very useful, they must be kept in a clean state, otherwise it will aggravate contamination of lubricants.

ROLE OF LUBRICANTS

To minimize friction in the mating parts, the use of lubricants is inevitable whether in the form of solid or fluid. It is generally observed that the proper use of lubricants prolongs the life of equipment and also minimizes unpredicted failures. Therefore, due care should be taken to periodically replenish them in concerned equipment. The working conditions also affect the quality of lubricants, so due care should be taken to assess their life as well.

In the systems where greases are used as lubricants on moving parts exposed to atmosphere, abrasive particles contaminating grease or oil increase the wear rate of the mating parts. Samples must be drawn, therefore, from the equipment at frequent intervals to check the quality of lubricants. The condition of equipment must also be monitored periodically as part of the maintenance schedule.

Due to environmental factors, chemical reaction takes place in the lubricants when in use on equipment and this may reduce the life of lubricants. The reactants may be in the form of oxygen, sulphur compounds, phosphorus compounds, chlorine compounds and organo-metallic compounds which may also react with metals. It is observed in practice that the environmental factors cited above could contaminate the lubricants faster, than when in actual use on an equipment.

HI-TECH OILS AND LUBRICANTS

With the help of advanced technology it has been possible to manufacture hi-tech mineral oils and lubricants, which can be used to reduce friction, wear, engine noise, fuel consumption and increase in power. The basic qualities associated with them include the following:

- ◆ High tech mineral based lubricants without any solid matter
- ◆ Based on formidable TEC 2000 mineral oil that ensures energy savings through reduced temperatures near immeasurable friction and significantly reduced wear
- ◆ Bonds oil molecules to metal surfaces ensuring immediate lubrication and protection of all moving parts

◆ Removes and counters carbon deposits

◆ Protects all metal from rust and corrosion

It is desired that the quality of the oils and lubricants in use must be examined at scheduled intervals to maintain their standard. If the same is not done effectively, will damage the parts and the operating costs also go high.

IMPACT OF ENVIRONMENT ON LUBRICANTS

Lubricants deteriorate in service by contamination or physical and chemical changes during the working. Contamination is the process, where the lubricants loose their properties and make them unfit for use. Such lubricants if used will reduce the operating life of the equipment and will enhance the operating and maintenance costs. The nature of contaminant will largely depend on the nature of products handled in the organization and also working environment. This process starts from the lubricants store itself. Improper handling of the oils and lubricants can be the main cause of contamination besides the impact of atmosphere in the form of dust.

For permissible contaminant concentration, no limit is set in advance and therefore, the equipment can be withdrawn from the production line when the limit exceeds the specified level. The problems caused by contaminants can be minimized by its regular/ periodic check up of its properties as prescribed by the manufacturer. Poor quality lubricants accelerate the deterioration process of machine parts and assemblies.

Various contaminant-monitoring techniques are available and must be made use of. The techniques include SOAP and Magnet Plug Inspection Analysis, Ferro-graph and X-Ray spectrometry etc. Adding a solvent to a lubricant specimen, centrifuging to remove clear layer from the top by evaporating the solvent, weighing the insoluble can be used for oil degradation analysis.

The presence of foreign solid substances in lubricants, can be monitored by withdrawing it from the system and carrying out physical comparison with standard specimen.

The core/earth faults in transformers can produce localized high temperature which can slowly breakdown the oil and produce ethylene gas. For this purpose, gas-in-oil analysis can be carried out to find out the faults. The insulating materials used in electrical systems may face the problems of gases, which arise from electric discharge or thermal breakdowns. This can be monitored with the help of gas-in-solid monitoring technique.

QUESTIONS

1. Discuss in detail the importance of lubricants in the context of the maintenance function.

2. Describe the various types of lubrication systems used in practice.

3. Indicate the factors which govern the selection of a lubricant for a particular application.

4. Discuss the importance of planning a lubrication schedule.

5. Discuss the need of proper handling and storing the lubricants. What precautions are necessary for storing lubricants?

6. How can a coding scheme help in better storing and handling of lubricants?

7. Discuss how lubricants should themselves be maintained.

8. Emphasize the importance and need of Hi-tech oils and lubricants.

9. Discuss briefly the importance of environmental impact on lubricants.

Chapter

15

Maintenance Material Planning and Control

INTRODUCTION

It is often seen in practice that many of the equipment/services remain unavailable for utilization for want of spare parts or because of non-availability of qualified craftsman. More often than not, such a state is usually due to poor planning and control of spare parts inventories as well as trained manpower. The importance of proper spare parts management cannot therefore be overemphasized for effective use of maintenance manpower and support systems. It is also known by experience that most of the breakdowns are repetitive in nature, which are unknown. During the purchase of equipment, a huge amount of money is spent on spare parts inventory, which can be better utilized through effective management and planning. The basic issues attached to the spare parts management are as follows:

- High financial commitment to the procurement of spare parts and their storage.
- Overstocking of spare parts due to unpredictable nature of failures as well as the concern to reduce the downtime of the equipment/facilities.
- Prolonged usage of equipment beyond its normal life cycle due to scarcity of funds as well as high requirements of production also increases the cost of spare parts inventory which is due to overstraining, wear and tear of the parts.
- Non-availability of proper type of spare parts due to obsolescence of equipment caused by advancement of technology.
- Non-interchangeability of spare parts due to different types of models and specifications pose problems to the users.
- Purchase of imported equipment also causes a lot of problems in procuring spare parts in a reasonable time, particularly for the components which have stipulated life.

◆ Political relations also affect the supply of spare parts in the case of critical in-service equipment procured from foreign countries by defence forces and other several industries.

◆ Government's import policy may also affect the supply of imported items.

◆ Provisioning of funds for stocking costly items may be a constraint.

◆ Communication gap between the concerned parties, i.e. suppliers, users and intermediate agencies may also cause some problems in achieving proper stocking of inventories.

PRESENT MATERIAL MANAGEMENT POLICIES

Through studies, it is learnt that most of the organizations in the country today do not have proper warehousing facilities for storing the items particularly the items which require a special type of atmosphere. For example, if the rubber items need to be stored in an airtight compartment, their storing elsewhere will affect their life and the equipment where the same items are used will not give proper service.

The technology transfer from the developed nations also causes the problem of interchangeability of items since the same are procured from different countries, although for the same type of industry. In many countries the practice of standardization of the items is not followed which poses serious problem.

Many organizations set up overhauling facilities, which require procuring the spare parts in advance since the date of overhauling is known. But for the first overhaul, the estimation of the exact number of spare parts is normally difficult. Sometimes, if a particular item is not available in the spare parts inventory, it has to be locally manufactured/fabricated. This also causes delay in completing the overhaul of the equipment.

In the developed countries due to fast pace of development and induction of new technology, the production of spare parts for the old equipment is not continued for long, since it is taken for granted that no spare parts would be needed and a new version of equipment will be produced soon. Also, the normal practice is to discard an item rather than to repair it, as the repair cost is some-times much higher than the cost of the new item. Therefore, it is of utmost importance for developing countries to procure a maximum number of spare parts as soon as possible after acquisition of critical equipment from the advanced countries.

Most organizations follow the A B C analysis for procurement of fast moving inventory items, but for slow moving and costly items, criticality analysis must be carried out. The concept of economic order quantities used by some organizations is sometimes purely of academic interest and does not serve the purpose in the desired manner. The overhead expenditure involved in the procurement of spare parts, which has an important role in the preparation of maintenance budget, is also not known to many organizations.

For the standardized equipment, the concept of spare parts bank is ideal and is being followed in many industrial organizations. The spare parts kept in a warehouse at a central place are supplied to the concerned units on demand. This also reduces the inventory cost, as each unit in the organization does not have to store the items. The following points need due consideration while in planning stage.

- The requirement of substitution of imported spare parts due to their general non-availability.
- Unwillingness of the concerned people to go along with import substitution.
- Non-availability of the complete specification data of imported spares.
- Lack of proper production facilities and materials for import substitution.
- Higher cost of indigenous production due to uneconomical quantities of spares.
- Contract conditions from the suppliers of imported equipment.

INFORMATION SYSTEM

An information system for spare parts to furnish the relevant information, if available at the user level, can help in procurement of spare parts at the appropriate time and at minimum cost. Properly maintained logbooks are very useful to know the past history of the equipment. It is seen that the maximum share of maintenance budget is consumed in procuring the spare parts. Keeping in view such problems, it is advisable to develop a spare part information system. Today all such information can be stored in computers and made use of as and when needed.

CLASSIFICATION OF SPARES

Every organization has its own way of classifying the spares, depending upon its requirements, into various categories such as critical spares, fast moving spares, slow moving spares, consumable and non-consumable spares. However, such classification should be based on the type of maintenance that has to be carried out, such as overhauling or preventive maintenance. Maintenance spares are those which are required during routine maintenance work such as seals, belts, fuses, bearings, etc. Overhauling spares are those which are required in a lot at the time of complete overhaul of the equipment.

INVENTORY COSTS

It is important to bear in mind all the cost elements associated with an inventory during its procurement and storing because it consumes a good amount of maintenance budget. The following main elements of inventory costs should be taken care of:

1. Capital cost involved in procuring the items.
2. Recurring storing cost involved in warehousing the items.
3. Loss due to non-usage of stored items, theft, and wastage, obsolescence of equipment due to change in technology.
4. Overhead expenditures in terms of payment of wages, rent, etc.
5. Insurance costs to protect the stored items against fire, theft, and other related risks.
6. In the case of computerized inventory, payments made towards the computer charges.
7. Stock-out charges incurred on account of emergency procurements at the time of actual maintenance work.
8. Penalty costs paid for not supplying a particular product in time due to production halts on account of breakdowns, which could not be repaired because of non-availability of spare parts.
9. Transportation cost for the spare parts, where the organizations are situated at remote places or the items are to be shifted from other stores.

SPARE PARTS COST OPTIMIZATION

The spare parts cost optimization is possible through many well-known techniques available, which include proper forecasting of inventory items, scientific analysis of maintenance activities by failure analysis and other methods to predict the failure time. Inventory control, refinement of purchase procedures to reduce the lead time, spare parts consumption control and standardization of products is also possible. The actual need of a particular spare part arises when it is required for replacement. If the same is available in stock, the machine downtime is minimized. The objective of an efficient inventory should be to keep the equipment downtime as low as possible; this will, in turn, also optimize the spare parts cost.

As indicated earlier, A B C analysis was in practice for inventory control along with VEIN–VED (vital, essential, important, normal–vital, essential, desirable) analysis. Looking into the limitations of the above concept, MUSIC-3D (multi unit spare inventory control-3 dimensional) approach came into existence. And it was realized that MUSIC-3D had numerous advantages, compared to others in the area of reduction of the spare parts cost.

It is of paramount importance that the spare parts supplied should be of high quality and reliability. The requirement of reliability becomes more pertinent in the case of high-precision equipment mainly used in defence forces and nuclear power plants.

Before actually accepting the spare parts, inspection must be carried out to see that the parts conform to the desired specifications with respect to accuracy and reliability. This will also reduce the chance of defective items being stored in inventory. The procedure of inspection will, however, depend upon the type of equipment and testing facilities available.

Factors Affecting Inventory Procurement

The purchase quantity and its time are important because any thing purchased extra would add to the inventory cost. The other parameters include the following:

1. *Economic parameters:* Here the purchase cost/manufacture cost of components/parts, its procurement cost, holding cost, stock-out costs etc. are important for consideration.

2. *Demand:* It can be deterministic or probabilistic and therefore due attention must be given while deciding the quantity.

3. *Ordering cycle:* It is identified by the time period between the successive placement of orders, which is possible through continuous monitoring of inventory.

4. *Lead time:* It is the time period between the placement of order and actual receipt of the items for use.

5. *Time horizon:* It is the time period for which inventory is to be controlled which is dependent on the type of inventory.

6. *Number of suppliers:* When the number of suppliers is low or a single vendor system is in practice the lead time could be high and therefore must be kept in mind.

7. *Number of items:* When the number of items is low the vendors will be reluctant to supply.

8. *Availability of items:* The most important aspects of inventory is its availability. If the particular item is not available in the market the same has to be manufactured at the local level.

SPARE PARTS PROVISIONING

The success of the maintenance department depends upon the timely availability of the right quantity and quality of spare parts. If proper record of maintenance activities is maintained, it will be possible to know the consumption pattern of spare parts for working out the provisioning quantities of the same for use in the future.

The basic objective of a maintenance function is to eliminate random failures or reduce their occurrence. If the preventive maintenance system is followed, then it is essential to know the number of components that need to be replaced during the maintenance process. Such quantities can be determined in advance for making arrangements for their procurement. However, in the case of random failures, it is difficult to predict the exact number and type of spare parts required, but by experience it can be known with greater accuracy. The tendency of the maintenance staff to store the maximum number of spare parts should be discouraged as it involves high costs. The number and type of spare parts required to be provisioned would depend upon the following considerations:

1. Criticality of the spare part, which will make the equipment totally inoperative or unsafe for use. This is a very important factor and should be given due attention.
2. Lead time required for procurement of each spare part.
3. Lifed components, which have stipulated life in terms of running hours or have life on the basis of calendar period. Such parts must be changed after expiry of their life period otherwise they may cause damage to equipment/system. No lifed component should remain in use after expiry of its life in the interest of safety of the equipment/system.
4. Cost is the primary factor for determining the actual type and the number of components/spare parts to be procured. It is unsafe to keep on reconditioning the equipment without caring for its material life. The costs saved in this manner may be harmful for the equipment/system.

For sophisticated and complex equipment the spare parts management is much more difficult. The problem further complicates in the case of imported equipment. To provision the right type of spare part in the right quantity and at the right time, requires considerable planning in terms of quality, quantity, price, material, and other parameters of delivery, transportation, etc.

When a spare part is to be ordered, the first thing to be finalized is its correct specification, which can be done only after having the complete details about equipment/system. If the item is procured from the original equipment manufacturer, then reference to the correct part number can give the relevant information about the component/item. However, if the possibilities of indigenous substitution must be explored in respect of imported spares, many more manufacturing details need to be available or may have to be developed through trials for satisfactory production of the substituted part.

Due attention should also be focused on the improvement of the spare parts to enhance the efficiency and life of the equipment. All relevant information available about a particular spare part should be passed on to the manufacturer for possible improvements in design and fabrication. For some of the items, suppliers may not agree to manufacture the item due to uneconomical quantity, then the same can be developed at the organizational level or through local manufacturers.

For smooth working of an organization, formal systems are essential and this concept also applies to the case of provisioning of spare parts. A proper system of ordering must be developed to reduce the lead-time and paperwork. The merits and demerits of the existing systems must be examined for effecting improvements.

STORES MANAGEMENT

The inventory of spare parts has to be stored in a warehouse. The management of the warehouse is an important function since it acts as a buffer between the user and the manufacturer(s). The main functions of managing a warehouse include:

- ◆ Receipt of orders
- ◆ Processing of orders
- ◆ Procurement of spare parts
- ◆ Inspection of spare parts
- ◆ Issue of spare parts
- ◆ Stock recordkeeping
- ◆ Valuation of stock
- ◆ Inventory control
- ◆ Stock verification at appropriate time
- ◆ Proper storing of spare parts
- ◆ Coding of spare parts
- ◆ Coordination with other departments/agencies
- ◆ Sorting out store problems

As large number of activities are thus associated with stores management. On the other hand, there are various factors, which influence the working of stores, such as warehouse location, availability of proper ventilation including air-conditioning, facilities installed for shifting heavy material, and procedures followed for preservation and safety of stores. Due care of the lifed items should be taken to avoid their wastage. Measures to protect stores against fire, corrosion, dust, and attack by insects, moisture, and sunlight, etc. should be duly implemented.

Stores management has to be applied across the board from the receipt of items to their proper delivery to the users. During the planning of the input process, the stores team has to perform various functions such as:

- ◆ Determination of proper quantities for provisioning
- ◆ Preparation of requisitions for provisioning
- ◆ Procurement of stores
- ◆ Quality assurance
- ◆ Coordination of all relevant work including documentation.

The above activities have to be systemized so that overcrowding of items is minimized. The items being received in the storeroom must be checked thoroughly for their quality, quantity, and expiry date of the parts, if any.

The output function of spare parts includes its issue to different departments or sales to the customers. The lifed items should be issued on first come first served basis. Whenever the stock of items falls below the pre-specified level, the procurement action must be initiated. Sometimes if an item is not available in stock, then the purchase department must initiate the required formalities on operational basis. Proper control of the indent quantity and issue will therefore help to minimize the storing cost as well as provide better information about future requirements.

The concept of multiple warehousing can also be implemented to cover large areas with an established distribution system. This practice is normally followed for public utility items and is more advantageous for costly items. These items can be stored in a central place and then supplied at the time of need. But for such a system, a well-established information system is desirable. Adopting this system can also minimize the cost of procurement of related items.

The objectives of a company dealing with the sales of spare parts must be to ensure availability of right quality of spares at the right time to the users with minimum investment at the seller's end. The credibility of the supplier will depend upon pricing of spares, efficient handling of customers' orders, and reliable delivery on schedule with sound after-sales service.

The important function of any organization is to keep proper and adequate quantity of spare parts in its stocks to carry out maintenance work efficiently in the stipulated scheduled time period and at minimum cost.

SPARE PARTS INVENTORY

For the purpose of maintenance work, the availability of spare parts in time is a must, otherwise the whole work might come to a standstill. With this in mind, the concept of inventory becomes essential for any organization having a variety of equipment. The main problem that arises is that of inventory or spares control, that is, how much to stock in stores as every procurement involves money and time. Where the equipment is imported, the problem of inventory becomes more acute as a good amount of time is involved in the procurement of the spare parts.

Normally every manufacturer of equipment provides certain items as spares during the sale of equipment and also recommends which items should be stored and in what quantity. Therefore it is the task of the maintenance group to properly estimate the number of spares required for the work. Various models are available which can be used to estimate the required number of spares. However in the case of bigger organizations spread over the entire country, a number of stores are required at different places with central store at the headquarters.

Forecasting the requirement of spare parts is the most difficult task for the maintenance department. However, the demand for spares depends on the factors enumerated below.

Failure Rate

It is important to know the failure rate of the components/parts over a specified period of time. This would help to know the number of spare parts required. Besides, the requirement of the spare parts also depends upon the following factors.

- ◆ Lead time needed to procure the items
- ◆ Availability of spare parts
- ◆ Number of equipment needing spare parts

Any effective material management system is based on the accurate forecast of the material demanded. In the case of maintenance work, this forecast is critical as the occurrence of failures is unknown and unpredictable. To actually find the requirement, the theory of probability of failures can be applied which will indicate the frequency of failures. If the population of equipment being used is high, this assessment will yield more accurate results otherwise it becomes hypothetical.

In the computation of spares requirement, two factors to be considered are:

- ◆ The cost of downtime of equipment/service
- ◆ The cost of stocking spare parts.

A balance between the above two factors has to be struck. In certain cases due to non-availability of items, a particular item may have to be stored on one-time basis though its use may be rare.

SPARES CONTROL

For the success of the maintenance function, the following important factors need to be looked into:

- Provision of accurate quantity of spare parts
- Reduction in the non-moving lot of spare parts
- Maintaining the optimum level of spare parts inventory

In practice, it is difficult to strike a proper balance between the above three factors, therefore, some controls are necessary for the same. The following are a few steps to minimize the range and scale of spares.

- Efficient indenting of spares
- Control over consumption
- Reconditioning/overhauling of used spares
- Use of management techniques
- Effective involvement of maintenance/operational personnel.

During the indenting process, the obsolescence of equipment and experience of usage of a particular system, must be borne in mind. During schedule maintenance work, effective control over consumption of spares must be exercised. The parts to be replaced must be checked before the incorporation of new parts. Sometimes it so happens that even though a particular subassembly has been changed during the operation of the equipment, the same becomes due during scheduled/planned maintenance work. This must be properly checked. At the same time, efforts must be made to reuse the item after reconditioning, but safety aspects of the system must be given due consideration.

QUESTIONS

1. Explain why it is essential to exercise planning and control of maintenance material.

2. Discuss the various maintenance material management policies being followed presently.

3. How can an information system be useful for the management of maintenance materials?

4. Discuss the importance of inventory costs in maintenance function.

5. How can the costs of maintenance materials be minimized? Explain.

6. List out the steps required for spare parts provisioning.

7. Describe how stores management can help in keeping a better spare parts inventory.

8. Discuss the factors, which need to be considered for implementation of an efficient spare parts control system.

Chapter

16

Decision Making in Maintenance

INTRODUCTION

Decision making is vital and important process for the success of any organization, which also applies to the case of maintenance. It is the choice out of several options made by the decision maker to achieve some pre-set objectives under the given circumstances. Decision making is a complex process in the higher management system. The complexity is the result of many factors, such as interrelationship among the experts or decision makers. The decision making process requires creativity, imagination and a deep understanding of the human behaviour.

Maintenance management involves critical decision making at various levels. Decisions related to operations and overall economic performance are vital for accomplishing the objectives of maintenance. The maintenance department of any industry works in association with other departments. Interaction with other departments must be well defined to avoid conflicts and to achieve smooth functioning of the system. Some of the important decision-making processes in the maintenance department are concerned with scheduling of routine servicings and inspections, minor repairs, temporary shutdown of a system for major overhaul, and repair and replacement of aging and obsolete equipment. Thus, it is essential for the maintenance manager to understand the basic methods of decision-making.

The following are some of the decisions required to be made by the concerned maintenance department:

1. Identifying the maintenance objectives and priorities.

2. Selecting the most suitable working method out of the available alternatives.

3. Maintaining the reliability and failure rate data of components/subassemblies/systems.

4. Predicting the impact of the external factors that are beyond the control of the decision-maker.

The flow chart in Figure 16.1 explains the decision-making procedure.

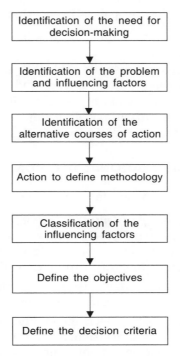

Figure 16.1 Steps in decision-making in maintenance.

The procedure outlined in Figure 16.1 can further be illustrated through an example. Suppose a failure of repetitive nature has been detected. The maintenance department needs to suggest a modification to minimize the failure frequency so that the remedial action is cost effective. The cost effectiveness can be judged on the basis of criteria of minimum cost of the modification. A possible course of action can be:

◆ Redesigning the equipment to avoid failure.

◆ Scheduling the replacement at fixed intervals to avoid actual failure.

◆ Replacing the part after actual failure.

The most suitable action under such situations depends on an understanding of the influencing factors. Such factors include causes of failure, cost of repairs, cost of redesign and benefits thereof. The attitude of the working personnel must also be given due weightage.

The safety of personnel also plays a vital role in decisions. It may be necessary to adopt a costly alternative to ensure the safety of both the operating and maintenance personnel.

Decision-making is a complex problem and some guidelines must be kept in mind while framing information systems for the maintenance function. The following points need attention:

◆ Information collected must be flexible to allow evaluation of many alternatives
◆ All alternatives must be able to accommodate changes because of diverse human nature and values
◆ They should be capable of being applied for analytical work and serve as intuitive models for evaluation keeping track of many alternatives and consequences.

PROBLEMS IN MAKING RATIONAL DECISIONS

The following are the problems faced by a decision-maker.

1. *Ascertaining the problem:* The problem must be defined in clear terms. As per Peter Drucker, 'the most common source of mistakes in the management decisions is the emphasis on finding the right answers rather than the right questions'.

2. *Insufficient knowledge:* It is important that the total information leading to complete knowledge is needed.

3. *Sufficient time:* The decision maker should have sufficient time to avoid hasty decision. Because decisions taken under pressure may not be rational and accurate.

4. *Environmental impact:* When the decisions are taken in haste may neglect the impact of environment and the decision may fail the test of optimality.

5. *Other factor:* The other factors may include, misjudgment of motives and values of people, poor communication, uncertaintities and risks and human behaviour.

The rational decision is ensured, if the process of decision-making is carried out systematically, whereby all aspects of decision-making discussed above are taken care of.

INFORMATION FOR DECISION MAKING

Efficiency of decision-making depends on the quality of the information based on which a decision is derived. Thus for effective maintenance management it is essential to design a system such that all the relevant information flows to the maintenance department. Such information must be processed with the right perspective prior to any decision-making. The maintenance department utilizes information for the following purposes:

Building-up Intelligence in Maintenance

Intelligent information concerning a system and its influencing factors can help the decision-maker to take an appropriate decision. In the case of any maintenance work, it is essential to understand the working environment and its effects on the system. Sometimes the failure caused may be purely due to environmental factors. Any modification of equipment without taking such factors into account may be a totally futile effort. To possess the basic facts, the maintenance manager needs to be aware of the related field data in order to verify the same through personal visits to the working site from time to time. Information about other similar types of systems in use elsewhere can also be collected. One method of building up intelligence in maintenance is to conduct detailed enquiry of every occurrence of failure and document it for future reference. Such documents should be readily made available to the personnel involved in maintenance decision-making. It is seen in practice that normally enquiries are conducted to pinpoint the reason for failure, which can be made known to all to avoid such mishappenings in the future.

Information for Design Phase

This phase involves designing of several possible alternatives to a problem. Here the required information has to be specially design-oriented in order to improve the existing system. The modern trend is to lay emphasis on the design of a system or component to ensure maintenance-free service for a prolonged period. To design such a system or to modify the design of an existing system, it is essential to have detailed information about the system or systems that are similar in nature. The maintenance department is normally the main source of information for such purposes. An information feedback system passes all performance and failure data to the design and planning department. Based on these data, a decision tree is derived to suggest possible safe and cost-effective alternatives. The information given by the maintenance department may be simple in nature, but crucial, for improving the performance of the system/equipment. A large volume of data having complex interrelationships needs to be processed prior to using them as the design inputs. Use of computers and intelligent software's, developed for the purpose, help in this regard.

Information for Selection of Alternatives

An information base is also necessary to select the best alternative.

Normally at the design phase, a number of alternatives are suggested. The final selection of a particular alternative course of action has to be made out of these available alternatives. Cost effectiveness is often used as criteria for such a selection.

The information collected must help the decision-maker in formulating a positive action. The choice may also depend upon the objectives involved. Sometimes, technical requirements may force an organization to select a costly alternative. Thus most of the decision situations are of multiple-criteria category that it further complicates the decision process.

The quality of any alternative decision depends on the quality of the input data from the preceding stages, i.e. intelligence and design phase. If a decision cannot be taken, the matter can be referred back to the design or intelligence stage. The organization at this stage should have complete information regarding the assumptions, resource requirements, and results of each decision structure. The use of simulation analysis may also help.

Information for Implementation Phase

The information collected in this phase can be utilized for implementing the decision. After selecting a viable alternative during the previous stage, a feedback system can be useful to make any alteration to the decision module. The information system must help the maintenance department in successful implementation of the decision. Reports can be prepared and circulated among the concerned personnel. The management tools like PERT and CPM make use of all of the relevant information to develop an effective course of decision implementation.

Information for Maintenance Phase

This information is most important since the total maintenance functions depend on it. The persons involved in the maintenance work should have a thorough knowledge of the working of the equipment/ system to understand the maintenance problems. The practice of *master of one* is often adopted as a better alternative to the *jack of all trades*. The practice of working on 'hit and trial' basis without proper knowledge of the subject often leads to wastage of time and money. Therefore, the maintenance work must be information based. For this purpose, all the relevant information about the maintenance function should be properly recorded and stored. These data should be used during analysis for preparing the maintenance schedules. The modern methods of maintenance should be used for improving the system/ equipment availability.

DECISION MODELS

As organizations are made up of individuals and as individual thinking and priorities vary from person to person, conflicts may arise in taking a decision. Therefore, it is desirable to build information systems to facilitate the decision-making process. There are various models of decision-making. Some of them are briefly mentioned below:

Rational Model

Some economists, mathematicians and management experts feel that decisions are made rationally because of the following reasons:

- The decision-makers know the objectives and their importance in the order of merit.
- They also bear in their mind the alternatives to each problem with their merits and demerits.
- They select the alternative which is likely to yield maximum profits and meets other objectives.
- They implement the selected alternative to yield the desired results.

However, in the case of the maintenance function, the alternatives are limited and the decision has to be taken purely on the importance and requirements of the equipment/system. The decision-maker does not know sometimes all the alternatives and their consequences. In spite of the limitations, rational models are powerful tools for decision making.

'Satisfying Approach' Model

Simon has proposed a model based on 'satisfying approach' rather than 'optimizing approach'. In this model the satisfaction factor is given much weight. To know how much satisfactory will a particular alternative be, extensive information about all the possible consequences of all the alternatives would need to be enlisted. It may not really be possible for the decision-maker to examine all the alternatives with their consequences as suggested in this model. Whenever possible, the new uncertain alternatives are avoided and testing rules are adopted. Some of the alternatives may be conflicting in nature. Therefore, an alternative which satisfies the functional and maintenance requirement should be selected.

Muddling Through Model

Lindblom (1959) suggested a radical deviation from the rational model of 'successive limited comparison'. In this case, values are chosen at the time when policies are framed. For instance, labour and management may not agree on values, but they can agree on policies. Because of the limits of the human rationality, Lindblom suggested 'incremental decision making' or choosing policies keeping in mind the comparison with the previous policies. Decision making is a continuous process in which final decisions are always modified to accommodate changing objectives. In the case of maintenance, the objectives do not change very much. Therefore, any decision has to be treated as final at the very outset.

LIMITATIONS OF FAILURE STATISTICS

There are various methods of failure analysis. However, a mere knowledge of these methods does not help if reliable data is not available. Acquisition of such reliable data for both human factors and machine factors is always a difficult task and demands proper planning and a systematic approach. With regard to human factors it is clear that great resistance exists at all levels for the collection of failure data. Many data collection methods have been used but were not successful because of lack of commitment shown by the top management. Some of the problems associated with the collection and analysis of failure statistics are as follows:

1. The motive for the data collection is not very clear to the person recording it.
2. Insufficient appreciation of the problems involved in the anyalsis of data and of the requirements to provide decision-making information to the right people at the right time.
3. Attitude of the data collector also play an important role during the process.

The data collection system should therefore be designed in such a way that it remains simple and does not miss out any vital piece of information. A simpler format must be developed to provide correct information at the appropriate time. It must also be ensured that the collected data are analyzed within the stipulated time. It may so happen that by the time the analysis is over, a number of replacements and repairs may have already been carried out and therefore the failure analysis would not bear any significant value.

INFORMATION COLLECTION

For decision-making, information is essential, but only that should be collected which is likely to be helpful in the decision-making process. Every decision would need a different type of information, hence due care should be taken during information collection. The information can be collected at different levels.

Departmental Level

Once the need for the information is recognized, it is easy to list out items of information to be used as database at the departmental level. The first information required is that of identifying the failure-prone components together with their frequencies. To optimize the maintenance cost it is required to know the repair cost of each component/ part including the material cost. Once the nature of failure is known it is easy to plan for manpower and material accordingly.

This information can be collected from the history cards of the equipment, which have to be maintained to carry out the planned maintenance work.

Organization Level

The information collected at the departmental level can be very useful either to decide the replacement policy of the equipment or to optimize the maintenance cost of the whole organization. The downtimes of all the equipment can be calculated to know the utilization of the plant/organization. The main work of this group is to collect the information from the respective departments and compile it.

In the case of the developing countries it is more significant to optimize the maintenance cost because of the high equipment replacement cost. For example, the expected life of the equipment in the developed countries is much less than that of in the developing countries. Due to the economic factors the equipment in the developed countries is replaced at shorter intervals, because of changes in technology or obsolescence.

For maintenance work, our own resources are utilized and job opportunities are also improved. This centralized information at the organization level can also be helpful to the manufacturers to improve or modify their products.

At the organizational level, maintenance information can also be useful to know manpower/spare parts utilization at the departmental level and also for re-planning the resources if required.

Public Enterprise Level

There can be public enterprises providing services of information processing for maintenance planning and implementation. The main aim of such an enterprise is to collect the information from the organization and compile it based on the type of equipment and its utility. These agencies receive the information in the form of reports forwarded by the client organization. The reports contain planned maintenance status and degree of its compliance as per the targeted goals.

The task of the information-receiving agency is to give the feedback by comparing the actual maintenance carried out with the scheduled maintenance work. This practice helps in making an estimate of the maintenance cost for comparison of the same with the production cost. The other utilization of this maintenance information is to help the upcoming industries to assess their status vis-a-vis established organizations.

USE OF POOLED INFORMATION

The information available at a central place can be useful at both the departmental level and the organizational level. For example, if a centralized inventory is maintained, the department or the organization can know the availability or non-availability of a particular spare part. Besides, other aggregate information can also be made available. The production cost for a particular item in the department may have a different value when the same is calculated at the organizational level. This fact will also be true in the case of maintenance function at both the departmental level and the organizational level.

When an up-to-date information system has been developed for the maintenance function, a new area of research in maintenance can be initiated, which will definitely help the organization. This is more relevant since the cost of equipment as well as the cost of maintenance is constantly increasing, particularly for imported items. Indigenous materials/processes have to be evolved to sustain the growing demand for new equipment/components.

R&D IN MAINTENANCE

Maintenance is associated with every equipment/service of technical nature and its importance increases with the technological complexity of the system. To reduce the maintenance time and therefore the downtime of equipment/facilities, it is essential to investigate the maintenance functions as a whole. In the past, this work was given the lowest priority, but now the concept of reliability and maintainability has gained greater significance. Therefore the need of research and development in maintenance has a greater role to play.

It is almost impracticable to manufacture hundred per cent reliable systems. Therefore, efforts are needed to minimize the failures. A highly reliable equipment may seldom fail, but its maintainability may be poor, i.e. it may take long time to repair the equipment. At the same time it is also not desirable to have equipment with poor reliability but with better maintainability. A compromise between the two has to be established depending upon the role of equipment. The third parameter associated with both is the cost, either of maintenance or that of design and manufacture.

The most widely used practice in maintenance had been that of breakdown maintenance, which cannot be dispensed with in any type of maintenance system being followed. Even with exhaustive scheduled maintenance, random breakdowns of equipment do occur.

For planned preventive maintenance, routines are prepared on the basis of past history of the equipment or components. Each important component is assigned a specific life period after which it has to be replaced, though it may still be in working condition. Compared to the breakdown maintenance, the cost of this type of maintenance is higher, but reliability is also high. In the case of corrective maintenance, guidelines are formulated for maintenance decisions so that the same type of failures do not occur. Sometimes, modifications in design and operation are also suggested.

The policy of fixed time replacement or repair, prior to failure, can be effective only when the failure mechanism of the component is mainly time dependent. Replacement may be cost effective if the part is likely to wear out over the life of the component and also if the total cost of such replacement is not high compared to the component cost. But to collect data for such items is difficult. However, this policy is not suitable for complex replaceable items, as more complex items do not exhibit a failure pattern that is time dependent. Secondly, the complex items are generally too expensive to replace or repair. For such items, the condition-based maintenance policy is most suitable.

In monitoring conditions, data are collected for corrective maintenance and/or for evaluating performance. However, there must be parameters specified that can be monitored to determine the deterioration level. Actions are taken on the basis of this monitoring to maximize the life of an item and to minimize the effects of failures. Of course, it should be noted that condition-based maintenance is costly in terms of time and money. It demands skilled personnel and state-of-the-art instrumentation.

The concept of opportunity maintenance is also practised for complex/continuously operating equipment because of high shutdown or unavailability costs. In this system, any available opportunity of system shutdown is used for maintenance of the whole system to avoid further failures.

Through the R&D work, alternative methods or modifications of used material can also be brought into practice. This work is more relevant in the case of imported items where, sometimes, the availability of the spare parts becomes very difficult due to obsolescence of the equipment. R&D is therefore essential to ensure the cost effectiveness of the adopted maintenance activity, in order to evolve better technical means and to enhance the life of components through modification of design or material. The phenomena of wear and friction in machines, and the development of most appropriate lubricating methods are also the areas of active research in maintenance engineering. For example, the use of radial tyres in automobiles has yielded better stability of vehicles, which has been only possible through exhaustive R&D work in this field.

BEHAVIOURAL CONCEPTS IN DEICISON MAKING

Every individual acts in a different way under a given situation. In a similar situation the decisions of two managers working in the same capacity may be different. This is for the reason that the decisions are influenced by personal factors such as behaviour even though the tools methodologies and procedures are same.

The managers differ in their approach towards decision making in the organization, for example one who looks for the completion of the work somehow is termed the achievement oriented. He will try to develop all possible alternatives, and therefore is more rational because of considering all the pros and cons properly and then conclude.

The manager's personal values will definitely influence the decision because of the conservative approach to avoid risk in all cases. On the other hand some of the managers would be willing to take certain amount of risk. Even though decision-making tools are available, the choice of tools may differ depending upon the motives of the manager. A rational decision in the normal course may turn out to be different on account of the motives of the manager.

The behaviour of the manager is also influenced by the position he holds in the organization. To keep up his image and future prospects he may not take any risky decision. The managerial behaviour is therefore complex in nature and depends upon the personal values, the atmosphere in the organization, the motives and the motivation and the resistance to change. Such a behaviour sometimes overrides the normal rational decision based on the business and economic principles.

It is observed that the rationale of the decision making will largely depend upon the individuals, their positions in the organization and their interrelationship with other managers. If the manager is averse to taking risk, he will make a decision, which will be subjectively rational as he would act with limited knowledge and also be influenced by risk averseness. Thus, it is clear that if the attitudes and the motives are not consistent, the decision-making process will be slow in the organization.

QUESTIONS

1. Why do decision making situation arises? Indicate the factors, which influence decision making in maintenance work.

2. Explain with the help of a flow chart, the various steps required in decision making in a maintenance organization.

3. How can information help in making appropriate decisions during maintenance work? Explain.

4. Describe the models used for decision-making in the context of maintenance functions.

5. Discuss the various levels of information collection for decision-making in maintenance work.

6. How can pooled information be useful in decision-making? Discuss.

7. List down the problems faced during decision-making.

8. What is the impact of individual behaviour on decision-making?

9. Explain the role of R & D in a maintenance organization.

Chapter

17

Environmental Impact on Maintenance

INTRODUCTION

The levels of maintenance and maintenance performance are both influenced by the environmental factors. Wear and lubrication are the primary concerns of the maintenance department. These two aspects depend on load, speed, and working temperature. The selection of lubricants is affected by the prevailing environmental conditions. The working environment also affects the type of maintenance to be adopted. Therefore, the maintenance practices vary from place to place, from industry to industry, and even from location to location of equipment in the same organization. The success of a maintenance system is largely dependent on the working conditions. The same piece of equipment may manifest quite different types of maintenance problems depending on the site conditions.

The adverse effect of environment on machines in petroleum, fertilizer, coal, and mining industries is more prominent because of the hostile working conditions. For example, it is seen in the chemical industry that the impellers of the centrifugal pumps need to be frequently replaced if they have to handle acidic water or slurry. The impact of moisture in steel structures and equipment can be seen and would require special attention during the maintenance work. The maintenance cost of such equipment increases due to the harsh working conditions.

The lubrication system of a machine often needs more attention of the maintenance department since the lubricants used, deteriorate with the passage of time. Their quantity also reduces during the use and therefore it must be topped up at regular intervals. The influence of the environmental factors on the lubrication system needs to be taken care of in greater detail with reference to the factors discussed in the following sections:

Technical Aspects

Lubricants are applied to reduce the frictional forces. The type of lubricant and the manner of its application are normally decided based on load and speed of the equipment. Sometimes under special conditions, a lubricant also works as a coolant and thus it forms a part of the environment. Loads and speeds may be fluctuating, uniform or intermittent in nature or may be a combination of these. The present trend is to design compact equipment capable of withstanding high loads and speeds. So the role of lubricants has become more critical and calls for their better quality.

Water used as coolant may affect the system if it is not of proper quality. The quality of water is dependent on the extent of the dissolved minerals, which varies with the nature and place of its source. Acidic elements in water may damage the whole system resulting in increase in the frequency of failures. For example, it is observed that the failure of radiators of dump trucks, employed in mines, increases during the summer compared to that in winter. This is attributed to the poor grade of water used in the radiators during that period.

The quality of lubricants deteriorates faster under adverse ambient conditions. For example, only the good quality lubricants can withstand the adverse effect arising due to the presence of water in the lubricant. Certain oils sometimes may cause problems by forming a stable emulsion within a short time. In such situations, a change in the viscosity of oil occurs which causes operational problems. Water effluence, therefore, is an environmental factor in lubrication systems.

In steel plants under certain processes the temperature of the system goes very high and this may affect the quality of lubricants being used. Here also, high quality lubricants must be used, otherwise it may lead to serious breakdowns.

Effect of Contaminants

Contaminated lubricants reduce the life of equipment/system as it carries certain amount of metal abrasives which are produced due to physical contact of two metallic parts. The nature of contaminants will largely depend on the nature of products/materials handled during plant operation and on working conditions. The contaminants may be classified as under:

- ◆ Solids, such as dirt, dust, abrasive material, metal wear particles, cotton fluffs, etc.
- ◆ Liquids, which can be soluble and insoluble acidic bodies being products of oxidation and improper combustion of fuels.
- ◆ Miscible liquids in the case of water contamination in bearings, or water and fuel dilution in internal combustion engines, may cause serious problems.

◆   Some lubricants get oxidized when in contact with free air, which is very common in any working environment.

To overcome the problems caused by contaminants, the lubricants must counteract the ill effects of contaminants. In practice, the oil/grease must be periodically checked/analyzed to assess its remaining life, which is normally specified by the manufacturers. As the working environment may contaminate the lubricant, the periodic assessment of its life at regular intervals is desirable. Attempts should, however, be made to minimize the degree of contaminants for a particular application so that the qualities of lubricants do not change.

Operational Factors

Certain operational factors also affect the lubricants during their usage. These include the following:

Process impact

In many cases the speed of machines is an important factor that decides the type of lubricant to be used. Any change in such parameters will affect the quality of the lubricant and therefore its life. The quality of lubricants is solely responsible for creating the correct thickness of film that is maintained between the two mating parts, which otherwise will overheat the system and may cause seizure. In rolling operation in steel plants where the equipment used is subjected to high temperatures, the lubricants which can sustain such high temperatures are used. For engines that are frequently started and stopped, there is a possibility of more carbon deposits inside the engine cylinder, which may necessitate scheduling of more frequent changes of lubricants as part of the maintenance routines.

Working conditions

The very purpose of selecting good quality lubricants is defeated if the actual working conditions are not taken into consideration. For example, in the case of dump trucks, especially in the open cast mines, if the haulage roads are not maintained properly, lubricants may deteriorate at a much faster rate. The road dust itself can cause a serious contamination of lubricants. The emission of toxic gases from the automobiles on the road will also affect the lubricants in a long run through chemical reaction. The environmental temperature will also have its adverse effect on the lubricants used in machineries and system.

Skill of operators

The skills of the operating personnel also play an important role as sudden braking and stoppages of the machine may cause overheating of the systems, leading to deterioration of the lubricants at a faster rate. One of the reasons of higher cost of maintenance is the inadequate technical know-how about the equipment and its systems by the operating and maintenance personnel. In many of the industries, people work without having any technical qualifications, which sometimes leads to serious lapses, resulting in accidents and damage to machines. It is observed in real life that a good skilled operator maintains his equipment better as compared to others.

The skilled and trained operators can take care of routine maintenance work, which are looked after by the maintenance personnel in a routine manner. With such people the number of breakdowns will come down because of regular and prompt checkups. The operator, who is responsible for operating a system, will take utmost care to keep it in a better condition. It is observed through experience that some of the operators create maintenance related problems deliberately.

Fire hazard

The petroleum products are always a source of fire hazards and a lot of loss is sometimes incurred owing to lapses on the part of maintenance personnel in the upkeep of the fire-fighting equipment. Care should therefore be taken in selecting the right types of hydraulic oils/lubricants for fire-fighting equipment. Fires are often caused by leakage of oils/lubricants through joints. Attempt should be made to use the fire resistant hydraulic oils/lubricants to minimize the risk. Wherever possible, the concept of temperature control may be applied to protect the equipment/system.

Temperature and Humidity

The selection of lubricants should be carried out with proper consideration of temperature and humidity of the working environment. The environmental factors such as temperature, humidity, light, and so forth affect the quality of the lubricants even during their storage. Therefore, lubricant storage must be designed considering the prevailing temperature and humidity conditions. The store-in-charge should be properly trained to maintain the quality of the lubricants, besides looking after the handling and storage facilities. The replacement of lubricants should also be planned based on actual working conditions rather than as a matter of standard practice. In India, weather conditions change from place to place with temperatures varying from one extreme to another and this condition is also responsible for high maintenance cost of some of the equipment.

Any equipment installed in a dusty or damp atmosphere is most likely to exhibit high frequency of failures and consequently the life of the equipment is also reduced compared to that operating under normal working conditions. Therefore, environmental factors play an important role in the selection and upkeep of the equipment. If the equipment selected does not match with the working conditions, the maintenance cost is bound to increase. Therefore, proper discretion should be exercised while purchasing the equipment/system keeping in mind the effect of the environment.

DESIGN IMPROVEMENTS

The effect of environmental factors can be reduced with the help of better design. For example, if the temperature rise is due to high speed the same can be controlled to save the equipment/system. If the parameters reach the highest value the system may tiff off automatically and thus avoid sudden breakdowns. For this purpose many equipment/systems are provided with alarms.

Material selection for the machinery used under adverse conditions also calls for proper attention. Under such conditions if the failure frequency is high, it immediately needs thorough investigation to ascertain the cause of failure. By simply changing the material the problem may be averted for example, impeller of pump used for pumping abrasive material.

The starting of internal combustion engines is difficult in cold areas therefore, necessary arrangements are provided in the form of heating chambers etc. to warm up the cylinder prior the start up. At the same time additives are mixed with cooling water to avoid their freezing in the said area.

Eco-friendly Machinery

To minimize the adverse affects either on the environment or from the environment of the existing equipment in use, the concept of eco-friendly systems is gaining popularity. Pollution of any kind let it be air, water or sound will leave its adverse affect on human beings as well as machinery around it. The use of electric trains in place of locomotives and CNG in place of petrol/diesel is one step forward towards eco-friendly systems. Introduction of multi-point fuel injection systems in internal combustion engines used in automobiles, is mainly to minimize the pollution and make the vehicles eco-friendly. The use of light indicators while driving on roads is to minimize sound pollution. At the same time use of better silencers is to keep the sound level at the lowest.

The introduction of water treatment plants not only helps in removing undesired elements from discharged water but also makes it reusable. The cow dung discharged by the animals can be used to generate power/gas, which can be used for minor works and waste can further be used for cooking.

It is observed that environment is affected by the pollution emitted from automotive vehicles, process and production industries. The efforts should be focused to prevent it rather than control it, which is the normal practice. In the subsequent paragraphs the importance of pollution prevention has been discussed.

POLLUTION PREVENTION

Pollution prevention is a way of looking at what causes waste and pollution and then figuring out the best way to reduce it before it is created. Here prevention means avoiding pollution at the source rather than controlling it after it is created. The various methods employed for controlling pollution are wastewater treatment, filtering air emissions, and landfills for solid waste. Attempt should be made to reduce waste first, and then recycle it. The approach of pollution prevention can save environment and save money at the same time.

Pollution is the waste which is not being used efficiently during the process. The prevention of pollution aims to identify the activities or areas in processes where wastes do not need to be created and other non-toxic material can be substituted. Pollution is not only expensive to treat, it also take energy and labour in the first place. In the long run, it costs less to avoid waste than to dispose it of. It is not practically possible to implement pollution prevention everywhere all at once. This must be done in the following steps (i) source reduction; (ii) environmentally sound recycling and (iii) as a last resort, treatment and disposal. Pollution control and treatment often moves the pollution from one medium (air, water, land) to another medium. For example, when wastewater from homes or factories is treated, the water is cleaner, but the factory is left with a sludge or byproduct that must be disposed of, often in landfill. Similarly other materials are moved from one place in the environment to other in different forms. In some cases, the volume or toxicity of the waste has been reduced. This pollution still causes environmental damage, takes lot of money, time and energy to deal with it. Pollution prevention aims to reduce the total amount of waste created before it has to be treated.

Emissions from fossil fuel power plants, automobiles, and strip mining and nuclear waste are pollution, which result from energy use. Pollution prevention also promotes increasing our reliance on renewable energy sources, such as solar, geothermal, and biofuels. It looks for ways to improve energy efficiency in every thing from heating and air-conditioning, to more efficient automobiles and electric motors and lighting. Energy efficiency reduces pollution to air, water, and land.

It is known that the nature is self-sustaining but human have chosen a non-sustainable approach. With the help of pollution prevention attempt is being made to make energy and recourse use more sustainable.

QUESTIONS

1. Discuss the impact of the environment on maintenance functions.

2. How do temperature and humidity affect the lubricants used in equipment?

3. Describe the important factors, which should be kept in mind while selecting the equipment for a particular environment.

4. Enlist the operational factors, which are responsible to maintain the quality of the lubricants.

5. Discuss, how design improvements can bring out environmental friendly equipment.

18

Manpower Planning for Maintenance

INTRODUCTION

The manpower resource is the most valuable and unique resource for an industry. The performance of manpower is unpredictable, which is not so in the case of other resources, such as machines. Unlike the other resources, labour is capable of exercising individual judgment and freely determining the course of its actions. Therefore, the manpower planning is one of the most critical jobs in an industry. Such planning is required to consider the industrial interests with special responsibility in a humanistic sense.

Manpower is the key economic resource which should be paid the same attention as is paid to finance, equipment, raw materials, production, sales, investments, profits, and maintenance functions. A company cannot hope to forecast its future manpower requirements accurately unless these are related to future production, sales, and maintenance requirements with its levels.

The rationale for manpower planning is related to the efficiency and effectiveness of the organization. Manpower planning is a process wherein the demand and supply of labour are equated. Profit maximization is often a goal of. an industry. To achieve this goal, capital and labour must be managed properly. Based on the price of labour and nature of the work, the industry must determine the requirement of manpower for the overall life of the industry. It is essential to find out what type of manpower will be recruited and at what time during the life of the industry. These aspects are the important considerations in manpower planning.

The need for manpower planning arises from many facets of the organization. It is unwise to assume that skilled workers will be freely available in the labour market whenever required, and so an adequate period of time must be set aside to organize and run appropriate programmes. With the workforce being maintained at an optimum size and mix of skills and experience, problems resulting from staff shortages or surpluses, production stoppages, promotions, inefficiencies and staff dissatisfaction may be alleviated.

However, it is not claimed that the planning process will always provide exact forecasts, and this is rarely achieved in practice since human beings display a commendable resistance to change. Rather it is suggested that there should be a more realistic expectation from the process, which should give general indications of the potential trouble spots. Such indications are invariably based on assumptions, these being implicit in the forecasting procedures that are used. Simulations of the possible outcomes arising from different assumptions about the factors and variables under study should provide a real aid to the manager. The inevitable slight inaccuracy in the forecast should not be taken as an indication that the process is a waste of time. Instead, it serves to emphasize the need for manpower planning to be a continuous process in which forecasts are checked against reality, and models and assumptions are constantly evaluated and updated.

The first step in the manpower planning process is to establish the planning horizon, i.e. the future period for which the manpower will apply. The aim being that the demand for labour services should remain matched to the supply of labour services over this period. The demand for labour is derived on the basis of the work requirements particularly on the frequency of failures in the case of maintenance functions. This can be done by considering the technical relationship between the input of labour services to a production process and the output of the commodities generated by the production process.

The current manpower stock is established, and estimates are made of the likely wastage or labour turnover during the forecast period. This wastage may then be subtracted from the current stocks to give an estimate of the supply of labour in the forecast period, and if necessary, an allowance may be made for normal recruitment during the period. The difference between the estimates of demand and supply of labour services is often referred to as the manpower gap, and one of the main components of manpower strategy is to formulate plans for closing this gap perhaps by recruitment and training or by planned redundancy. The period over which forecasts are made will inevitably vary between organizations, depending on such factors as the training time required for new employees, the time period over which new capital equipment is developed and installed and the stability of the environment in which the organization operates.

The models used in planning should explicitly contain variables relating to the rate at which labour is utilized, the factors determining wastage or characteristics of the turnover process, and the factors influencing the supply and demand for labour. Manpower planning also needs to consider inputs like the learning process and the characteristics of the system of industrial relations. Economics, sociology, psychology and other applied sciences all have a part to play in a good manpower model. The influence of the environmental factors should not be overlooked during the manpower planning.

OBJECTIVES OF MANPOWER PLANNING

A manpower planner's function in a maintenance system as part of an overall planning team is:

- ◆ To interpret maintenance forecasts in terms of manpower requirements for all the related activites of the maintenance functions which are unpredictable in nature.
- ◆ To indicate manpower constraints as a result of the company's policy for the future, i.e. expansion or retrenchment.
- ◆ To formulate the organizational policies to retain the skilled and trained manpower.
- ◆ To maximize manpower retentively ratio with better perks, facilities and promotional avenues.

The acute shortages, particularly of skilled technical and qualified manpower in many companies, have brought home the need to anticipate manpower requirements a few years ahead if they are to ensure an adequate supply of qualified manpower. It is again important to see that manpower resources are utilized as effectively as possible. Manpower planning may be conceived as a strategy for the acquisition, utilization, improvement and preservation of the company's manpower resources. It is often observed that qualified and trained personnel change their organizations for better promotional avenues.

STAGES OF MANPOWER PLANNING

Manpower planning for maintenance needs should be carried out following a systematic methodology. Such planning involves the following stages:

- ◆ Defining the overall maintenance objectives for the stated period ahead
- ◆ Converting these maintenance objectives into manpower objectives for the same period, allowing for changes in the maintenance system, number of equipment, and so on
- ◆ Designing a manpower information system to obtain and process data in the most efficient and economical way

◆ Undertaking a manpower inventory in terms of composition, deployment and utilization by tabulation and careful analysis of the current manpower resources within the system

◆ Analyzing manpower requirements or the demand forecast based upon the maintenance objectives. This requires:

(i) A decision concerning as to where the maintenance system or the workshop management should concentrate its efforts

(ii) An assessment of the overall size of the maintenance system

(iii) Estimates of the resources needed by the system

(iv) Formulation of advance maintenance plans

(v) Detailed targets for each management level.

To forecast the relationship between the workload of the maintenance plan and the manpower required and analysis of the past manpower performance and trends in the adoption of maintenance practices are needed. The forecast must also provide the type of manpower required.

Analysis of Manpower Supply

A vital part of the manpower planner's work is to indicate the constraints on policy imposed by conditions prevailing in the local and occasionally the national labour market. The planner needs to consider the internal supply of labour already in the organization. Labour turnover also affects the internal supply of manpower as do promotions and transfers, and consequently, these too will have to be analyzed for adequate manpower planning. The requirement of manpower may also change due to advancement/upgradation of technology and this should also be taken into consideration.

Improving Manpower Utilization

After hiring or procuring manpower, the management must consider the ways and means of improving the quality of manpower utilization. Manpower utilization is measured in terms of achieving targets with improved productivity. Consequently, the determination of efficiency indicators and measures of productivity enhancement must be incorporated in the manpower management system. When these have been established, then shop management can set targets that are achievable and can be monitored. It is possible to indicate the achievable targets by comparing the present performance with the trends in the past in the same shop floor or at other places of work.

In the light of these, the recruitment policies must be examined and developed. This helps in formulating policies required to improve manpower management, particularly with respect to absenteeism, illness, accidents, and disputes etc.

Identifying Training Needs and Assessing Training Effectiveness

As indicated above, the advancement/upgradation of technology demands that the needs of training of the existing personnel be properly identified. The development of manpower and the training scheme should therefore be planned on the basis of the future needs for trained manpower. Consequently, the first step is to identify the training needs within the maintenance system, and the next step is to assess the effectiveness of such training policies and methods. Both these jobs are grossly dependent on the nature of the industry and must be carried out following a methodology suitable for the situation.

Controlling Manpower Costs

A systematic appraisal of manpower costs gives a better appreciation of the contribution and value of the various manpower activities such as recruitment, training, and development. It also quantitatively appraises such problems as absenteeism, sickness, and labour turnover. This will assist in making cost-benefit analysis and in deciding on the approach, which will give the best subsequent yields.

TIMESCALE OF MANPOWER PLANNING

Manpower plans are basically of two types — short-range and long-range. The short-range manpower plans are detailed and from them, stem factual plans for recruitment, promotions, transfers, training, etc. The long-range plans, however, are much more concerned with the strategic problems and indicate the overall direction of manpower policies. As such, their use lies in their value to shop management in deciding the maintenance policy.

Consequently, the lines of approach are different for the short-range and long-range manpower planning. For the long-range plans, the manpower planner, on the basis of analysis, has to translate the maintenance objectives into a forecast. He can then discuss such a forecast with shop management in order to incorporate their assessment of the situation.

However, with short-range plans, the emphasis is reversed in that the shop managers make their own forecasts based on their intimate knowledge of the situation. This forecasting should be supported by the data and analysis of the manpower planner. The long-range plans must be sufficiently flexible to cater to changing circumstances.

Forecasts of both demand and supply may turn out to be wide off the mark, and the main assumptions on which they are based may therefore require substantial revisions. To make provisions for these, two procedures are normally followed. First, the manpower planner must closely monitor the analysis and forecast by comparing them with what actually happens on the ground. This should provide an early warning of the sort of adjustments needed and at the same time, enable the manpower planner to test and improve the analysis and forecasting methods that were used. Secondly, there should be some system of reviewing the manpower development by workshop management at regular intervals. The following factors are important in the contemplation of manpower planning in any organization.

Demand and Supply of Labour

The technical relationship between the input of labour services to a production process and the output of commodities generated by the production process is depicted by the production function. Variations in the flow of output are partially determined in the production function by various inflows of the input in labour services over a particular period of time. Manhours worked, should be used as an indicator of labour input since it takes into account the variations in the number of hours worked per employee and thus more accurately reflects the flow of labour services.

The problems of labour supply are multidimensional. They concern the decisions of individuals and form the financial and social constraints faced by both the individuals and the organizations. The individual is the cornerstone of supply analysis, as individual decisions in aggregate create the supply stock, which confronts and interacts with the job offers of firms.

Age Distribution of Workforce

The age distribution of workforce is important to ensure that the age structure is not unbalanced in a particular sector or category of employees. It might be discovered, for example, that there are too many employees of the same age in a particular department or occupation and steps have to be taken to avoid promotional difficulties, which may arise as a result. The age pattern is also useful to determine the effect of a policy of enforced retirement at any selected age for any particular trade or department, and also to determine the cost implications of any such policies.

Length of Service

This gives a breakdown of the labour forces by period of service. The particular value of a periodic analysis of this nature is that it highlights the departments or occupations in which it is difficult to get the labour to settle. It will also show the size of the hard core that is loyal to the company and thus help to improve the effectiveness of recruitment procedures and devise measures to curb labour turnover.

Geographical Locations

For skilled and qualified personnel, the geographical location of the organization is very crucial. It is seen in practice that people are reluctant to go to the places, which have poor communication facilities. An analysis of where employees live is also useful when considering the location of the new work sites or possible movement of labour. If such developments are long term, an age limit should be established so that the older workers due for retirement in the intervening period are not included in the analysis.

Labour Turnover

The extent of voluntary and involuntary terminations in a company is important to know not only for the purposes of manpower supply, but also as an indication of the general health of the organization. Compilation of the wastage statistics, even if done thoroughly, is of little value unless the figures are interpreted to discover the real cause of labour turnover and action taken accordingly to control wastage.

Analysis of labour turnover must be carried out in detail. The total wastage rate alone will not indicate whether or not a situation is satisfactory. Even if the total number of persons leaving an organiz-ation is by itself not considered excessive, a more detailed breakdown may reveal the areas that are unsatisfactory. A statistical breakdown of labour turnover by each department in itself may indicate the source of the problem.

It is also important to identify the type of employees in *wastage analysis* to see whether or not a lot of key people in any function are being lost. A further breakdown by grade will show at what level people are being lost. What action should be taken to equalize these losses will be for the personnel department to decide, depending on just how large the talent drain is. If it can be shown that the company is consistently losing talent over a period of years, it is best to start with the rectification of the company's recruitment policy.

Nature of Job

In India some of the jobs are undertaken by certain group of people and others are not allowed for social reasons. But in the developing nations such practice is not accepted and everybody has freedom to select the jobs of their choice. To certain jobs traditional values are also attached which again do not permit others to be part of it.

Reasons for Leaving

An analysis of the reasons given for terminations will indicate the steps needed to reduce turnover and its effects and whether or not the cost of any measures to reduce turnover can be justified. It is likely that some simple measures can remedy a difficult situation. Furthermore, such an analysis will indicate ways to improve the management—employees relationship as well as job-effectiveness. The reasons for leaving can be broken down under the following headings:

- ◆ Economic factors which affect the labour market and impinge on the work situation such as the level of employment.
- ◆ Factors within the organizations, which affect the employees.
- ◆ Factors concerning the employee as an individual or factors affecting a group of employees.
- ◆ Factors related to the working conditions.
- ◆ Factors related to the society/social structure of the working force.

The reasons for leaving will vary from one type of individual to another. Opportunities for promotion, work culture, lack of independence and interruptions in the workflow, size of the work group are some of the reasons given by people for leaving the organization. Location of the organizations may be another reason for leaving a particular job. It is normally observed in the Indian context that people prefer their native place even though better opportunities are available outside.

Analysis of External Supply

If a company is considering an expansion of its workforce, then together with an analysis of the internal supply situation, the manpower planner will have to analyze the external manpower supply as well. This helps to find out whether or not enough manpower with the required skill is available locally or whether a different labour market would need to be accessed.

MANPOWER FOR MAINTENANCE SYSTEMS

The factors that must be considered while planning manpower for maintenance systems are enumerated here, taking the example of the mining industry.

Type of Operation

The manpower planning of a mine maintenance system will vary depending on whether it is an underground mining or an open-pit mining operation. Generally, the number of equipment employed in the latter case is higher. Opencast mining deploys a large number of technologically advanced machinery. Therefore, the maintenance system must be well planned with the required level of skills and the requisite number of such skilled maintenance personnel.

Moreover, the maintenance site should also be properly selected for easy transportation of the faulty equipment. Based on the conditions of the working site and the type of machinery used, specialist manpower would need to be developed.

Extent of Mechanization

The extent of mechanization of the concerned mine will decide the manpower strength of the maintenance system. The mechanization level will dictate the skill of the maintenance personnel to be used in the workshop. Special training needs are associated with complex mechanization. In case a mine opts for a modernization programme, motivating and training the existing maintenance manpower for new challenges of maintenance operations may have to be looked into very carefully. The attitude of the working personnel is important to accept new and advanced technologies in interest of their organization, people and nation as a whole.

Type of Mined Mineral

The type of ore or mineral has a bearing on deciding the strength of the maintenance manpower. The working conditions in a mine are often adverse and very challenging, which result in frequent breakdowns demanding more care from the maintenance personnel. These aspects must also be considered in deciding the manpower requirement in a mine maintenance system.

Number of Shifts

In case of productive systems it is desired that it should work round the clock in the form of shifts. The number of maintenance personnel required depends on the number of shifts practised. A three-shift working needs more maintenance manpower than a double-shift working. In a three-shift working, operation-crews are also trained to cope with emergencies and first-aid maintenance work. Under such working conditions equipment/machineries are provided to the maintenance personnel for their jobs of maintenance and other two shifts are used for production.

Size of Mine

The size of the mine is determined by the targeted output capacity.

This is one of the most prominent factors directly related to the strength of the manpower required for mine maintenance. The more the output, the more is the number of machines employed for production. This invariably requires more maintenance personnel as well to keep them in normal operating conditions.

System of Mine Maintenance

The manpower requirement for maintenance also depends on the system of maintenance being followed. The manpower requirement is different for preventive maintenance, breakdown maintenance and condition-based maintenance. There is a difference in the level of sophistication or skill required on the part of the maintenance personnel under different circumstances. The strength of manpower required will be the highest for preventive maintenance and the lowest for the breakdown maintenance for a particular type of equipment.

Strategic Location

The strategic location of the mine in relation to manpower supply also affects the manpower strength. If the availability of skilled/trained personnel is low in the adjoining areas, then the requisite manpower may have to be recruited from outside the working area with additional incentives. Under such a situation, the manpower turnover has to be studied carefully in order to understand the factors, which would help minimize the turnover of the manpower.

MANPOWER ESTIMATION MODE

In the organizations where planned maintenance systems are being practised estimation of maintenance manpower is easy and can be calculated by using models. When the developed schedules are not followed properly it will pose problems due to backlog of the work. By knowing the total number of equipment likely to visit maintenance shop, it will be possible to specify requirements of skilled, semiskilled and helper etc. with the help of the model.

In practice, however, any planned schedule is subjected to change for practical reasons. Under such situations it is advised to keep marginal allowance in the schedule during the planning stage. The amount of allowance is based on the past experience and the working environment of the organization. The use of dynamic scheduling can take care of such exigencies. For the development of any model assumptions are must which include the following:

- ◆ The arrival pattern of the machine is known in advance
- ◆ The percentage spillover is known

♦ The conditions of manpower idle and machine idle are known

♦ The planned manpower is always available

Considering a fixed number of days in a week, schedule can be drawn for the arrival pattern of the machines to the workshop. If the work on any machine is not completed in a week, it is then completed in the next week. Therefore, model is developed with varying spillover.

Keeping the cost parameters, optimum number of men for the work is calculated week wise with due care of spillover. It is observed that the crew size calculated using the model yields schedule with minimum cost. Sometimes, due to the non-availability of spare parts, the planning is to be redone to calculate the spillover. As the maintenance load is probabilistic in nature workforce hiring and firing must be permitted. However, resource reallocation is possible month wise.

Effective Manpower Utilization

It is seen that the nature of manpower requirements for maintenance is unpredictable; this can be improved by using the following suggestions.

1. The maintenance planner must understand the needs and objectives of the maintenance department with the work requirements.

2. The top management must extend full support for manpower planning for maintenance functions.

3. The complete and authentic information in respect of maintenance functions must be recorded efficiently.

4. The planning of maintenance manpower should be done under long-range to take care of unforeseen problems.

5. The planning techniques used should match the available data and degree of accuracy required.

6. The information collected in any form must be updated and revised as the experience in the field is gained.

7. The equipment due for any type of maintenance must be made available to the maintenance department well in time with proper support facilities.

QUESTIONS

1. State the need of manpower planning for maintenance functions.

2. How can the forecasting techniques help in manpower planning? Explain.

3. Briefly describe the objectives of manpower planning.

4. Discuss the stages to be followed in manpower planning.

5. State how can the manpower planning be made more effective.

6. Discuss the factors, which influence the manpower planning for maintenance work.

7. Discuss how maintenance manpower can be estimated using estimation models.

8. Explain the measures taken to effectively utilize the maintenance manpower.

Chapter

19

Energy Conservation and Maintenance

INTRODUCTION

Friction and wear are the two important factors associated with any moving part of the machine/equipment. No motion is possible without friction, but at the same time a higher value of friction consumes more power or fuel. It is well known that when the piston rings of an internal combustion engine get worn out, the efficiency of the engine goes down and it consumes more fuel. Wherever there is friction, wear of the component becomes an integral part of the system. Higher frictional resistance is associated with the rise in temperature of the involved system and increased wear rate. All the basic elements of machines such as bearings, gears, valves, cylinder and piston require lubrication to reduce friction and wear. Wear reduction through proper lubrication helps in saving energy. Wear studies through condition-monitoring techniques also help in reducing the unscheduled breakdowns. Implementing preventive maintenance and replacing the worn-out parts before they fail, can also make the system/machinery energy efficient.

SOME DESIGN FACTORS FOR ENERGY CONSERVATION

Due to the high cost of petroleum products, it is desirable to select fuel-efficient equipment/machines. The earlier concept of giving much value to design for the sake of safety was at the cost of consumption of high values of energy. With the development of new techniques it has become possible to implement the design without causing any risk to the safety of the equipment or personnel.

To save energy in a hydraulic system, for example, attempt should be made to control the temperature as effectively as possible to minimize the loss of energy due to overheating of the components/parts. In certain cases where the load is supported on hydraulic pressure, the running of the pump over long time periods may cause excessive heat, but the energy can be saved by the use of an accumulator, or a variable volume pump. The variable displacement pump with appropriate controls can be used in place of the fixed displacement pump to minimize the power losses. Besides the use of variable displacement pumps for conserving energy, sometimes the hydraulic circuit can also be made more efficient by the use of different valves. One example is the use of solenoid-operated relief valve in the place of the ordinary relief valve. These valves are normally used when fluid is to be supplied to several systems at different maximum working pressures.

While selecting drives for operating machines it is important to know the actual power requirements. It is generally seen that designers select oversize drives, firstly for the sake of safety, and secondly for future expansion work. High-power drives consume more fuel or electricity, which can be minimized through proper selection of their size.

Due to increased pace of mechanization the demand for electrical power has become excessively high, and the need of the hour is to pay serious attention to the problem of electrical energy conservation. It is therefore essential to identify the areas where it is possible to conserve electrical energy. One effective measure can be the planning of proper sites for the substation/transformers to minimize the distribution losses. The proper selection of energy efficient drives can also be a step towards minimization of losses in the overall system. It is important to use high efficiency machines for a continuous operation, even though the initial cost of the machines may be high. The cost can get compensated in due course of time by way of improved utilization/efficiency.

The other way to conserve energy can be through the improvement of energy efficiency by implementation of the following short-term measures:

- The unnecessary idle running of electric motors and equipment should be minimized.
- The induction motors should not run at full load as their efficiency decreases with decrease in the output load. Under the fluctuating load conditions, the hydraulic motors provide better efficiency.
- The capacities of the equipment working together in system must be matched so that all equipment operate at their rated capacities.
- Proper use of compressed air can also sometimes lead to significant energy savings. The air losses in pipelines and valve joints should be reduced as far as possible. Moisture in

air does not directly hamper energy savings, but moist air causes leaks and thus damages the equipment.

◆ The transformers must be placed near the load end to decrease the transmission losses.

◆ The selection of materials for fabricating different parts should be given due care since weight leads to higher power consumption. The bigger structures should be made from light material, wherever possible.

◆ High efficiency motors which may be initially costly but are energy efficient, should be used as far as possible.

Long-term Energy Conservation Methods

These measures include major equipment design changes or implementation of new techniques, which not only give direct energy savings but also provide cost effective utilization of equipment.

It is the general view of the experts that the energy consumption in the industrial sector in our country can be reduced by about 20% if proper measures are taken. In this regard, the training of personnel also plays a very significant role towards improvement of the performance of equipment and machinery. Skilled and trained personnel can be an asset to an organization in the attainment of defined goals in the area of energy conservation.

Besides the above, the following can also be considered as the means for the conservation of energy in their respective fields.

Thermal plants: For the production of electricity, one of the main sources is the thermal power plant. In these plants the working fluid is steam. To improve the efficiency of the thermal plants, the quality of the steam has to be improved, which can be done with the help of economizer, pre-heater, evaporator and superheaters. To improve their performance further, the following measures can be effectively employed.

(a) Use of binary vapour cycle, since a single fluid does not meet all the working requirements.

(b) Use of coupled cycles where two Rankine cycles have been used in series.

(c) Use of combined cycle plants—a higher energy conversion efficiency from fuel to electricity can be achieved, since the combined power plants operate through a higher temperature range. The combined plants may be of the following types:

● Gas turbine–steam turbine plant

● MHD–steam plant

● Thermionic–steam plant

● Thermoelectric–steam plant

It is observed that with the application of the above methods, a substantial amount of energy saving can be achieved with the reduction in fuel consumption.

Automobiles: The multi-point fuel injection (MPFI) system has been successfully installed in automobiles to achieve fuel economy. The advantages of this system are improved combustion of fuel, uniform distribution, quick throttle response, easy cold starting, high volumetric efficiency and minimization of fuel condensation in the chamber.

Energy management: With shrinking fuel availability, the need to conserve natural resources is of paramount importance. The following methods of energy management must therefore be adopted.

(a) To meet the peak load demand, interconnect power networks that might have different demands on them.

(b) Use the most efficient power plants for base load power generation and the old, less efficient power plants to meet power generation requirements during peak periods.

(c) Install the smaller low capital cost power plants, for example, gas turbines and hydroelectric plants.

(d) Incorporate energy storage systems. The energy produced using the various sources can be stored when it is surplus and released whenever required to meet peak loads. This concept of energy storage is new but will have applications in future.

ROLE OF SAFETY ENGINEERING IN MAINTENANCE

Because of accelerating technology and demands for a 'safe operation' a need has arisen to formally organize safety efforts throughout the system's life cycle and therefore, the concept of System Safety Engineering has been developed. The objective of system safety engineering is to safely integrate all system components in a manner consistent with other system which involves the application of scientific and management principles for the timely recognition, evaluation and control of hazards. The systems safety analysis methods provide a proactive approach to analyze systems for potential hazards that may threaten the health and safety of workers.

It is learnt from history that industrial safety consists of three "Es"—Engineering, Enforcement and Education. Application of system's engineering has brought in drastic changes in safety aspects of equipment/system. The techniques used include risk management, fault tree analysis and others. Enforcement has been facilitated by the legal requirements to comply with codes, standards and regulations that are constantly being updated. Education through task related training is a cornerstone of safety training.

In the case of minor injuries such as damage of fingers or back straining it has important role to play. The more promising techniques here include back strengthening exercises and proper use of tools. Protective gloves and belts also offer the potential reduction in the rate of accidents. Proper planning and written instructions do help to minimize the accidents in the area of maintenance. Preventive maintenance practices also help in minimizing the rate of accidents indirectly because the number of failures would be minimum.

The term hazard is used to describe an unsafe situation in a workplace. An accident is a realization of a hazard. Improvement or elimination of these hazardous conditions is essential for increasing the safety of workers. It is further observed that engineering solutions to accidents and system approach to accident prevention have reduced the number of accidents significantly.

QUESTIONS

1. Name a few major measures, which can minimize energy losses in hydraulic and electrical systems.

2. Enumerate some short-term measures, which the maintenance personnel can implement to achieve energy conservation.

3. Briefly explain how energy can be saved in internal combustion engines.

4. Name the methods that can be employed to save energy in modern thermal power plants.

5. Explain the concept of energy storage.

6. Discuss the steps required to minimize the rate of accidents in maintenance activities.

20

Maintenance of Mechanical and Electrical Systems

INTRODUCTION

In most of the industrial organizations the maintenance work is mainly oriented towards ensuring the best working conditions for the mechanical and electrical systems. The main job of such maintenance is to reduce the wear rate of the mating surfaces and to eliminate unpredicted and accidental damage or breakage. Proper and timely attention to a problematic component results in enhancement of the equipment life. Maintenance of a mechanical system also involves the application of condition monitoring of its components and taking adequate measures as soon as deterioration or deviation from the normal signature is noticed.

Maintenance aspects of some most commonly used components of the mechanical and electrical systems are briefly discussed in this chapter.

BEARINGS

Bearings are used to support the load. Normal functioning of all rotating machines depends on the good condition of the bearings. Bearings can perform satisfactorily, provided adequate lubrication is designed to reduce the friction and wear of the mating parts. It is essential to consider proper clearance during the assembly of the bearings so that adequate lubrication can be provided. The right kind and proper quality of the lubricant in sufficient quantity must be applied at specified intervals. For improved life, the lubricant must be free from contaminants like metallic particles and acidic organic substances. In no case the bearings should be allowed to run at a temperature that is beyond the specified limit. Such high temperatures may cause material disorder. A well-established rule to check the proper functioning of the bearing is that while running, the housing should be warm but not so hot as to be unbearable. There should be suitable arrangements to monitor the bearing temperature using the appropriate temperature gauges. In the case of critical equipment, a warning system should be incorporated. One common reason of high running temperature of the bearings can be the inadequate clearance between the shaft and the housing. A bearing operating with inadequate lubrication will show heat discolouration and metal smearing, particularly at the contact surfaces.

In case any abnormality is found in bearings, the machine should be shutdown for investigation. In the pressurized lubrication systems, the normal operating pressure must be monitored from time to time to know the condition of the bearings. The excess lubrication of the bearings may also cause problems.

When the machinery is overhauled, care should be taken to use replacement bearings of the same clearance as originally furnished with the equipment. A misalignment of shaft or housing will cause the individual balls in the bearing to follow an uneven path in the inner or outer ring raceways. The skidding across the race groove and in the cage-pocket will cause excessive heat and high temperature, which may result in cage distress. The ball and roller bearings should be handled with the greatest possible care to avoid mechanical abuse and damage due to corrosion. The bearings should be constantly protected from all forms of dirt or foreign matter that might dent or wear out the highly polished surfaces of the balls, rollers, and races.

The other important problems associated with bearings are the noise and vibration of the bearings. Alignments of the shafts and mountings of the bearings should be properly implemented to minimize or eliminate vibrations.

Care should also be taken in selecting a proper bearing lubricant to ensure that the oil or grease possesses the required qualities to protect the bearings for the full range of operating temperatures. If the bearings are properly protected from the foreign materials, the rate of wear and tear of the bearing parts would be minimized. Basically, the lubricant is supplied to a roller bearing to support the sliding contact which exists between the retainer and the other bearing parts and to accommodate the sliding that is unavoidable in the contact area between the rolling element and the raceways. It also protects the highly finished surfaces of bearings from corrosion and partly transfers the heat from the bearing surfaces.

When grease is used in bearings as the lubricant, proper care should be exercised to remove the old grease from the bearings before the application of the fresh grease. Excessive grease should not be applied to the moving parts as it may not facilitate better working of the system; instead, it may create problems of attracting some foreign material. The lubricants must also be protected from moisture.

Wherever necessary, water repellent lubricants must be used. To keep the bearings in proper order when the machine is kept idle for a long period, they should be cleaned and packed with anti-rust agents according to the instructions given by the manufacturers.

The unused bearings should be placed in a suitable container having a clean, cold petroleum solvent or kerosene and allowed to soak for 12 hours. For cleaning the bearings, transformer oil, spindle oil or automotive flushing oil should be used. If the bearing is to be re-lubricated with grease, some of the fresh grease may be forced through the bearing to remove any remaining contamination. For the assembly or disassembly of bearings, recommended tools must only be used to avoid damage to any part. However, if they are not readily available, a mild steel bar may be used taking proper precautions against damage to any of the bearing parts. Whenever difficulty is faced in removing the bearings with a puller, dry ice can be used to shrink the shaft and hot oil is then poured in the housing to expand the bearing. In emergency, steam can also be used in place of the hot oil but subsequent care should be taken to clean the bearing in order to remove the moisture.

FRICTION CLUTCHES

Friction clutches are used for connecting and disconnecting power from the driver to the driven unit in motion. They are required either for the purpose of starting a machine, or in those cases when a machine has to be run without the load. Friction clutches are most commonly used in internal combustion engines. The main problem associated with clutches is their overheating during use and the wear of the contact surfaces. Proper lubrication can be used to minimize the wear. If overheating is noticed in bearings even after lubrication, the clutch should be dismantled and the bearing checked for the presence of foreign material in between the bearing surfaces. In general, the clutches must be cleaned and reassembled with fresh grease at least once in every two years of use. Sometimes the heat generated may be high when the clutches are required to connect a machine to a heavy load. Air cooling of the clutches can be one way to minimize the overheating.

COUPLINGS

The basic function of any type of coupling is to transmit the torque from the driving to the driven shaft and also to offset the mis-alignment of the shafts. Proper selection of couplings for the specific requirement is very essential for trouble-free operation. Normally, couplings are selected depending on the required speed and torque characteristics. The installation of couplings must be done with care to ensure proper alignment, as misalignment would increase the rate of wear and shorten the service life. Misalignment may also induce undue vibrations of the coupled items, affecting the overall performance of the system.

For proper working of a coupling, it is important to check its temperature at specified intervals for detecting overheating. Fluid couplings are provided with fusible plugs to protect them from overheating. For high speed machines, dynamically balanced couplings reduce maintenance problems.

FASTENING DEVICES

For the transmission of power mechanically, the use of belts, pulleys, and fasteners is very common. The selection of belts involves checking for proper material and construction of the belts and also the environmental and mechanical conditions involved. Depending upon the power transmission requirements, the width and thickness are decided. The metallic fasteners should be made up of the proper material and they should be constructed for specific working conditions and load requirements. Proper crowning should also be provided.

Periodic inspection should be carried out to check slackness, breakage, slip, and other faults in belts. Though the repair requirements for the belts are less, proper tension is necessary to achieve the desired power transmission. Alignment of the two pulleys carrying the belts is also important to avoid slip. Belts cannot give good service if the proper alignment of the pulleys is not maintained. If the tension of the belts goes down the pre-specified limit, dressing of the belts may be necessary. Whenever leather belts are used, extra care has to be taken to maintain them by proper dressing/lubrication. This is essential because the heat generated during use, makes the belts dry. Any extra oil or grease on the belts thrown from the machine bearings should be cleaned. The condition of the belt plies must also be checked from time to time to assess the life of the belts. When rubber belts are used, they should be dressed periodically with vegetable caster or linseed oil. This will soften the belt surface and provide adhesiveness between the belt and the pulley face.

Transmission shafts normally fail due to abnormal deflection or torsional fatigue. The deflection of a long shaft can be observed through visual inspection. In such cases, proper support to shafts should be provided to minimize the deflection.

CHAINS

Chains are also used for power transmission in some of the industrial applications. Proper chain tension is essential because a tight chain will exert excessive pressure on the bearings, and a loose chain will provide noisy operation. This may also lead to faster wear either on the chain rollers or sprockets. Lubrication is required to minimize metal-to-metal bearing contact at all points.

Every chain drive should be checked for alignment periodically. It should also be checked for slackness. In some cases the chain length can be reduced by removal of the one or more links and whenever this is not possible the chain should be replaced. A worn-out chain or a chain with damaged sprockets should not be used. A new drive must be checked frequently for possible interference of the two bearing surfaces.

If the chain drive is exposed to the atmosphere, the grease applied should be replaced periodically to minimize foreign materials in between the two contact surfaces. The quality of the lubricants must be checked before use. The chain drives should be provided with covers to eliminate the entry of dirt and dust, whenever possible.

GEAR DRIVES

Gears are widely used for power transmission particularly where changes in speed and torque are desired. Gears can also be used to change the direction of rotation from the driver to the driven shaft. The advantages of using gears for drives include:

- ◆ Economy of operation
- ◆ Adaptability
- ◆ Long-service life
- ◆ Positive drive
- ◆ Safety of operation
- ◆ Minimum maintenance requirements.

The satisfactory performance of the gear drives depends upon their proper design and manufacture and the place of application. Installation of gear drives is also an important parameter for achieving good service. During installation, the following points must be considered:

- ◆ The gear drives must be properly supported to prevent misalignment of gears.
- ◆ Good quality couplings should be used to couple the shafts of the driving and the driven units.
- ◆ Wherever possible, torque unit switches must be used to minimize the overloading and jamming of the drive unit.
- ◆ The gear drive unit should not be operated where the oil temperature exceeds the values recommended by the manufacturers.
- ◆ When forced-feed lubrication systems are used, they should be checked to observe, if the oil is being pumped properly. The pressures must be checked at the specified time intervals. The relief valves must also be adjusted at specified time intervals.

The lubricating oils to be used for the gear drives should be of high grade, high quality, and within the recommended viscosity range. They should also be anti-corrosive and should not possess foaming properties. When the gear drives attain high temperatures due to their speed, the lubricant used should have good resistance to oxidation. Heavy-duty lubricants must be used when the gear drives are subjected to shock loading or heavy duty. To prolong the life of the bearing, it is also essential to apply sufficient grease which would form a thin film over the rollers and races of the bearing. Due care should be taken to prevent the entry of foreign materials inside the gear drives. Lubricants used in gear drives must be protected from moisture, dust, and chemical fumes, which otherwise would deteriorate the quality of the lubricants.

Periodically the oil level must be checked for any leakage of oil. If any unusual noise is noticed during the inspection, the unit must be shutdown until the cause of the noise has been determined and rectified. Temperature monitoring of the gear drives is also essential to maintain the quality of the lubricants. To prevent rusting of the gears and bearings, they must be run every week when the equipment is kept idle for the longer periods.

The main cause of gear drive failure is the wear of the gears or the breakage of a tooth due to shock loading. Wear can be minimized by proper lubrication and filtering of the oil periodically. Filtering will remove the metal particles which otherwise aggravate the wear rate. Tooth breakage may also occur due to gear tooth deterioration.

SUPPORT EQUIPMENT

Any mechanized organization has to be supported by ancillary equipment. The maintenance of such support equipment/facilities itself plays a vital role towards the performance of the organization and therefore this aspect should not be overlooked. In the present-day organizations, air-conditioning is very common and is provided to improve the performance of men and machines. By control of temperature, humidity, and purity of air, the actual performance of any working system can be improved and certain types of failures such as bearings, gear boxes, clutches, etc. can be minimized. The condition of the lubricating oils can be better maintained when the system is working under an air-conditioned environment. The other effects such as moisture, dirt and dust can also be controlled which normally deteriorate the lubricating oils.

Air-conditioning equipment can be divided into two groups: the first being the air-handling equipment and the other one being the refrigeration equipment. The air-handling equipment includes fan, filter, heater, air inlet, and outlets. The refrigeration unit includes the compressor and its drives, the cooling coil and the condenser. For efficient working of all these components, due attention should be paid to their maintenance work. The maintenance aspects of some of the important components are described below:

Air Compressors

Most of the compressors use the splash type lubrication system and therefore the oil level must be checked periodically as the extra oil in the tank may cause carryover of the oil into the system which is bad for all the control devices. At the same time, the low level of the oil in the tank will not provide sufficient lubrication. All parts of the compressor should be cleaned at regular intervals including the filter for better performance. The air-handling units must be checked periodically for any leakage or damage to pipelines to avoid accident.

Centrifugal Compressors

These compressors are used in large-size refrigeration plants. For proper functioning of the compressor, it is necessary to check the oil level of the lubrication system at pre-specified time intervals, including the oil temperature and pressure. The oil must be replaced once in a year because of its contamination over a period of time. All joints must be checked for refrigerant/water and air leakage as specified by the manufacturer.

All cooling and heating coils must be checked for tightness and cleanliness. Here tightness means that leakage through joints must be eliminated, and cleanliness refers to the cleaning of all pipelines internally and externally to remove foreign particles, which are likely to be deposited during the operation of the compressor.

Air-Cooled Condensers

For removing heat from the refrigerants, air-cooled compressors are used. Condensers must be inspected periodically for physical damage of the coils. Fan blades should be checked for alignment and clearance. The condenser must be cleaned at regular intervals. Motor must also be cleaned from outside. For the purpose of lubrication, the instructions from the manufacturers must be followed strictly. For better maintenance function, a logbook must be maintained equipment-wise. All major operations of the maintenance must be recorded in the logbook.

COOLING TOWERS

These are used for heat rejection from the cooling medium. Their size depends upon the requirement of use. To protect the frame, casing, etc. from corrosion, the cooling tower should be painted at regular intervals. Gear-driven fans must be checked for alignment to meet the given specifications. The oil level should be checked weekly and be replaced once in a year. The water distribution system must be checked for proper distribution and also for any obstructions in the passage. The water basin should be cleaned to remove sludge formation, which may otherwise restrict the water flow. Algae growth should be cleaned from all the parts of the tower and water should be treated to prevent the growth of algae.

Dampers

Dampers are used for proper regulation of air in the air-conditioning plant. During their maintenance, all moving parts must be checked for free movement at regular intervals. Blades should be checked and be replaced if damaged. They should be properly adjusted for efficient functioning. All drives must be checked periodically to prevent any tension and damage to the fibre bolts. Particular care has to be taken to replace the V-belts as and when required in order to ensure an uninterrupted power transmission.

Fans

Fans are used to deliver air at the required capacity and operating pressure. Care should be taken to remove the fine dust particles from the fan blades, which otherwise may even cause unbalance. Bearings should also be checked as per their time schedule. Alignment of the fan and the motor is essential for proper working of the fan units. All filters used in the conditioning system should be cleaned at regular intervals. Filter elements should be taken out and washed. In some cases, the filter elements are replaced as per the schedule.

Heat Pumps

These are used to extract heat from the system or to add heat to the system, i.e. they can be used for heating as well as for cooling. For successful operation of the heat pumps, the following points must be borne in mind.

1. The outdoor coil must be checked and cleaned at regular intervals to remove the frost which is formed because of low temperature. If defrosting is not done properly, the efficiency of the system will fall.
2. As per the time schedule, the filters must be cleaned/replaced.
3. Any dust on the surface must also be cleaned at regular intervals for efficient functioning of the heat pump.

4. Other components of the system must also be checked/ cleaned as per the routine recommended by the manufacturer.

STEAM TURBINES

Steam turbines are used to produce electric power in almost all countries and are used beyond their intended lifetimes because of power requirements. Opening up of machines is expensive therefore; sufficient information must be collected while making such a decision. The efficiency of the steam turbines reduces because of deposits on the blades and erosion of internal clearance, which can be detected and monitored, using condition monitoring by performance analysis.

Vibration analysis can detect the occurrence of shaft rubbing and other conditions on rotor line, but cannot detect internal wear or deposition. On the other hand, performance analysis can be applied to find increased fuel consumption, or in reduced output, or both. However, it is preferable to try and determine the internal condition by testing, and use this information to help make the decision as to the extent of overhaul. The main wear out problems with the steam turbine are the following:

◆ *Blading:* It is erosion by solid particles or by water droplets which some times breaks the part and can be detected by performance analysis.

◆ *Bearings:* It is damage to white metal and can be detected with the help of performance analysis and vibration analysis.

◆ *Valve spindles:* Here loading of turbine tends to have a blade-washing effect this can also be detected by performance analysis.

◆ *Generator, rotor, stator:* It normally have insulation faults, which can be detected by electrical plant testing.

◆ *Condenser:* There is a possibility of air leakage and tube fouling which can be found by performance analysis.

◆ *Feed water heaters:* Here also there is a possibility of air leakage, tube fouling by scale or oil which can be monitored with the help of performance analysis.

◆ *Valve HP, IP bypass:* These parts have problem of leakage and can be detected by acoustic leakage detection and performance analysis.

ELECTRICAL EQUIPMENT

The operating temperature of any electrical machine affects its life expectancy, along with a number of other factors affecting the machine rating. The type of service to which a machine is subjected also affects its life. The dynamos that are required for intermittent duty have smaller frame size than those rated for continuous duty. The larger the frame size, the greater is the surface area and better is the heat dissipation.

Dynamos that are not enclosed completely and are well cooled by fans, have higher output rating than those which have completely enclosed frames of the same size. Similarly, the high speed dynamos have higher ratings than the low speed dynamos of the same frame size.

The type of insulation used in stator and rotor coils not only affects the allowable temperature rise but also the life expectancy of the dynamo. The factors that affect the life of insulation are sustained heating at an excessive temperature, mechanical shock and vibration, moisture, corrosion, atmospheric conditions and/or extremely low service temperatures. At low temperatures insulation may shrink, become brittle and crack as a result of excessive vibrations.

Most of the electrical machines are specified by the parameters such as, allowable temperature rise; duty cycle; rated voltage; current and speed; frequency; and output (kW or kVA) capacity. Some other details are also printed on the nameplate of the machine.

Motors rated on the basis of a 40°C ambient temperature, should be capable of continuous operation at rated load without excessive damage to the insulation provided:

◆ The ambient temperature does not exceed 40°C.

◆ The altitude does not exceed 1000 m.

◆ The atmospheric conditions (dust, moisture fumes) do not inhibit proper and adequate ventilation.

◆ Applied voltage does not exceed ± 10 per cent above or below rated value.

◆ The applied frequency does not vary ± 5 per cent above or below rated value.

◆ All mounting methods are being followed as specified by the manufacturers.

Insulation Temperature

Studies have shown that for every 10°C increase in operating temperature of a continuously operating motor over the recommended hottest spot temperature limit, the winding life is cut to half. Conversely for every 10°C reduction in motor operating temperature under the rated limit, the winding life is double.

For different type of insulating materials used in practice, the manufacturers normally specify the temperature limits.

Frame Size

In general, for the same horsepower rating; as the rated speed of the motor is increased the frame size is decreased. In short, high-speed of the same horsepower are always smaller than low-speed motors. The same relation applies to DC generators and AC alternators.

Standard Ratings

The standard voltage ratings for DC generators vary from 125 V–600 V and for DC motors from 90 V–550 V. However, AC single-phase motors have ratings from 115 V–440 V and AC polyphase motors from 115 V–6600 V. The AC alternators have voltage ratings from 120 V–6600 V.

Effect of Duty Cycle

The various types of duty cycle used for electric machinery include, continuous duty, intermittent duty, periodic duty or varying duty.

For the same hp or kVA rating capacity, the continuous duty machine will be larger in size physically than the intermittent duty machine. The larger size results from conductors of larger diameter and heavier insulation. The duty cycle is closely related to the temperature, and therefore, is generally taken to include environmental factors also.

Maintenance of Electrical Machine

It is observed in practice that preventive maintenance and routine inspection techniques conserve and prolong the life of the machines. An induction-type machine requires only periodic lubrication, whereas a machine with self-lubricating "lifetime" bearings requires no lubrication. However, the initial installation of electrical machine is very important so far the life of the bearings is concerned.

Machines, which are provided with brushes such as dynamos, need proper cleaning of the brushes at regular intervals in addition to lubrication. High speed series wound (DC, AC or universal) motors should not be selected for long and continuous duty cycles because of brush-sparking which requires frequent replacement.

Excessive oiling of electric machinery is as damaging as underoiling or insufficient lubrication. Oil leakage onto stator may cause insulation breakdown of AC and DC stator windings. The machine equipped with centrifugal switches will need more attention as compared to other machines. If a centrifugal switch mechanism is "stuck in its running position" the motor fails to start. On the other hand if the switch is "stuck in starting position" the starting winding overheats and the motor fails to reach rated speed.

It is observed through experience that rewinding of integral-horsepower motors is less expensive as compared to fractional-horsepower motors. In some cases replacement is less expensive as compared to rewinding. Consequently, small motors are generally replaced. Human senses are of immense help to maintain any equipment and therefore visual inspection may reveal a number of troubles. For example, noisy motor is an indication of worn bearings, overloading, or single phasing. A burnt odour, indicates overheating/burning of insulation. An overheated bearing or winding is detected by touch.

The maintenance of the some of the important electrical equipment is discussed in the following section. This only highlights the general precautions to be taken during the scheduled maintenance work.

Electrical Motors

The problems associated with motors include the following:

- ◆ Starting of motor
- ◆ Will not start even at no load, but will run in either direction when started manually
- ◆ Starts but heats rapidly
- ◆ Sluggish-sparks severely at the brushes
- ◆ Abnormally high speed—sparks severely at the brushes
- ◆ Reduction in power—motor gets too hot
- ◆ Motor blows fuse, or will not stop when the switch is turned to off position
- ◆ Jerky operation—severe vibration.

For the above problems the following must be checked:

- ◆ Open in connection to line; open circuit in motor winding; contacts of centrifugal switch not closed or starting winding open
- ◆ Contacts of centrifugal switch not closed; starting winding open
- ◆ Centrifugal starting switch not opening; winding short-circuited or grounded
- ◆ High mica between commuter bars; dirty commuters; worn brushes; open circuit; oil soaked brushes
- ◆ Open circuit in shunt winding
- ◆ Winding short-circuited or grounded; sticky or tight bearings; interference between stationary and rotating members
- ◆ Winding short-circuited or grounded; grounded near switch end of the winding
- ◆ High mica between commuter bars; dirty commuters; worn brushes; open circuit; oil soaked brushes, shorted or grounded armature winding.

Transformers

The transformers used in practice are of two types, i.e. liquid filled or dry type. The function of the both types of the transformers is the very same, to step-up or step-down the voltage. It is observed in practice that their failures are mainly due to improper maintenance. However a correct interpretation of maintenance data from transformers is vital for increased reliability, long life and advanced information on possible need of replacement. This information is provided with a warning of approaching service problems for which corrective actions can be taken in time. If the following points are taken care of in time through scheduled maintenance it will help in minimizing the failure frequency of the transformers.

Variation in sound level

If the noticeable change in sound level is detected, first check the input or output voltage. It is observed that an increase in voltage increases the sound. When the measured voltage is within the specified limits then check the windings for internal damage. An increase in sound level can also be the result of load current distorted by harmonics.

Tank heating

When hotspots on the tank surfaces of liquid-filled transformers or enclosures of dry-types, are severe enough to blister or discolour the paint, it may indicate the existence of open or shorted internal load connections. If this is noticed the transformer must be deenergized as soon as possible and electrical tests must be performed. Winding resistance and impedance measurements are especially important when tank heating is observed.

QUESTIONS

1. Discuss the important maintenance aspects of the following mechanical components: (a) Bearings (b) Friction clutches (c) Couplings (d) Belts (e) Chains (f) Gear drives.

2. What is the role of ancillary equipment/facilities in an organization? How should such facilities be maintained? Give your answer with particular reference to air-conditioning equipment, compressors, and cooling towers.

3. Discuss the important points, which must be kept in mind while maintaining electrical equipment.

21

Advances in Maintenance

INTRODUCTION

The recent past has witnessed enormous growth in the area of material science and technological upgradation. As a result, the complex and sophisticated equipment/systems have been developed and put into use. The cost of such equipment/system is bound to be high. Hence effective utilization of equipment is of utmost importance for overall economic viability of the system. To keep the system availability high, the maintenance engineer has the responsibility to provide the requisite support system. All types of maintenance systems discussed in previous chapters will not yield satisfactory results if the maintenance support system is weak/inadequate. The modern machineries and systems need well-trained and qualified persons to handle them, be it operation or maintenance.

Today, maintenance and infrastructure management activities have great economic value in at least two ways. First, they provide value to organization because maintenance and infrastructure management activities may be characterized as high value added, while at the same time requiring low value investments. In the recently industrialized countries, careful development, in harmony with environment and energy requirements is essential and must be undertaken through well-planned maintenance and infrastructure management activities.

The second way in which maintenance and infrastructure management activities provide economic value is through job creation. With the creation of maintenance and infrastructure management, companies move towards outsourcing territory activities. In effect, the expansion of territory sector has enabled the enterprises that perform maintenance and/or infrastructure management services for many different companies to realize important scale economies, improve the level of personal utilization and obtain greater uniformity in employment levels, service quality etc.

The advances in the area of maintenance and infrastructure facilities have been discussed in the following section.

RELIABILITY AND MAINTENANCE

Reliability engineering provides mathematics, statistics, and analytical instruments to understand the logic of a technical system's operations, including the manner in which they breakdown, and estimates of the probable requirements for repair. The fundamental role of this science in maintenance is therefore self-evident. Knowledge of reliability engineering of a given system, nevertheless, is closely related to the availability of historical data relative to its characteristics breakdowns, or the breakdown characteristics of the family of identical production systems, in relation to the environment and condition of use. The availability of this data is generally limited, due to lack of culture relative to the management and processing of data, which is generally available in the organization, but inadequately managed.

The basic concepts of reliability and its calculation are discussed in Chapter 8 with suitable examples.

To assess the reliability of a complex system/equipment it is desired to collect the operational data which can be done with the help of accelerated tests to save time.

TELEMATIC MAINTENANCE SERVICES

Maintenance was originally understood as a prevalently manual activity, characterized primarily by repairs, replacement and reconstruction. The need to guarantee a greater degree of dependability to particularly delicate and critical systems, such as aeronautic and nuclear systems, has given strong impetus to probability studies relative to breakdowns, imposing the development of preventive maintenance to avoid occurrence of breakdowns based on conditions, predictions and prognoses.

At the same time new information and communication technologies have made it possible to implement maintenance and inspection action from a distance, further enriching the meaning of maintenance, which has already become an inspection service to improve possible prevention of breakdowns (with a view to economy, safety, the availability of system etc.) and is increasingly and rightfully seen today as a service to improve management of assets with reference to entire life cycle, including the phase of disposal, which is particularly important for environmental purposes, with the prospect of virtuous recovery of reusable materials.

The increasingly complex nature of maintenance, which has come about as a consequence of the more complex systems in use, together with increasingly binding requirements for preservation, have definitively mark the birth of the science of maintenance.

In fact maintenance is no longer a question of making repairs, or at most, defining more or less articulated programmes of prevention. On the contrary, it has today acquired the meaning of complex management service, oriented towards the prevention of breakdowns, but also towards pursuing a variety of other objectives, ordered differently on the basis of company's requirements and its reference environment, which may also be commercial in nature.

The tele-maintenance systems are based on the centralized management of information collected in the field and transferred over the Internet and, in their general operational logic, they permit a very general applicability, once the necessary network of sensors and transducers has been designed. In this context it is interesting to cite the prototype of tele-maintenance system perfected by the Department of Mechanics and Aeronautics at the 'La Sapienza' University of Rome (2007).

This prototype is based on a logic of management of geographical information and allows the intelligent processing of data with a view to making predictions (through neural software) and has been experimented on various types of equipment (elevator plants, acclemization systems, medical dialysis equipment).

In the field of maintenance, for example, nano-technologies can be effectively applied to improve the durability of materials, through the deposit of suitable lining, where pure materials may not yield the desired results.

DECISION SUPPORT SYSTEM BASED ON ARTIFICIAL INTELLIGENCE

This system uses the concept of artificial intelligence which can analyze the data monitored through sensors, placed on the components/subsystems. The choice of the faults which will be monitored is fundamental, because monitoring of all faults will be costly and may not be useful in due course of time. Skimming off the most relevant faults is thus necessary to obtain precise diagnostic system while keeping the leanest structure possible. In the next step it is important to identify smallest possible subset of signals which allows to effectively detect and recognize each of the impending faults selected for monitoring.

There are several reasons behind the choice of using soft computing and artificial intelligence techniques instead of adopting a mathematical approach. First of all it is difficult to satisfy the necessary conditions of physical systems/equipment in mathematical formulation. Secondly, what traditionally made human experts, much more apt for diagnosis than machines, are their flexibility and ability to deal with the situations highly classifiable with certainty.

In order to develop a diagnostic expert system, the rule base, representing codified diagnostic knowledge, must be created and validated through a test session. In order to obtain rule base, the proposed methodology suggests the use of neuro-fuzzy system capable of extracting rules from experimental data (self-training process).

Once the expert system has been codified and validated, the last step, concerning the creation of the maintenance management module can take place.

When the diagnostic module is ready for operation, a second module is necessary to plan the maintenance actions maximizing the economic efficiency of each mission. In order to create such a module, different analysis have to be done. First of all, the characteristics of the product must be closely considered in order to understand how to gain the maximum benefit from the single maintenance mission. The cost of each component, for example, can be extremely important in deciding whether to change a subsystem or not with still relatively short residual life when a mission is planned to maintain a product in the same geographical area.

Secondly the maintenance process must be analyzed and considered as its most important feature. The maintenance management module must mirror the structure of the process in order to be effectively used in practice. Finally, the maintenance management module can be implemented to get a tool capable of considering a number of installed products at the same time of planning the mission on the estimated ageing values and matching the needs of the single company reflecting the structure of its own maintenance process.

The neuro-fuzzy system proposed to extract knowledge from data proved to be a powerful procedure to generate a set of fuzzy rules describing the deteriorating behaviour of a product of interest. The software used can generate the rule base in a very short computational time and without requiring high computational resources. Further, two more modules have been designed for the maintenance module: the spare parts manager and the maintenance mission manager. The first one allows to visualize the estimated number of components needed for a certain period of time defined by the user. The value is calculated starting from the expected failure dates of the monitored products. The second one is designed to obtain a better scheduling of maintenance missions, optimizing the number of maintenance actions in the same mission, either changing more than a component on the same product, or scheduling actions on products in the same area when the expected residual life does not economically justify another maintenance mission in the near future.

USE OF RADIO FREQUENCY IDENTIFICATION (RFID)

The two key aspects mentioned in this section can be identified as the factors, driving the adoption of RFID in maintenance:

Drivers from the perspective of maintenance (e.g. in transparency of maintenance process, costs of maintenance), and technical innovations in information technology (e.g. mobile terminals, tablet computers, wireless communication, component miniaturization, embedded systems with sensors), the use of distributed information carriers and mobile terminals has been rationalizing the maintenance work in the companies. Leaps in technical performance, technologies, mobility and on-line communication are enhancing the quality of maintenance technologies.

A transponder: A word combining transmitted respond; also called RFID tag or level. It is a device for receiving a radio signal and transmitting a different signal automatically.

A transceiver unit: It is a read-write device with integrated antenna. A transponder has an integrated circuit, to store data that is connected with antenna. The chip either extracts its operating power from the radio frequency signal emitted by the radar (passive transponder) or a self-contained power source in the form of battery (active transponder).

Variants of RFID used in maintenance: A wide variety of demands is made on technical concepts for the use of transponder in maintenance. These stem from the following:

◆ The operating conditions to which manufacturing equipment can be subjected, e.g. dirt, dust, extreme temperature, aggressive media, metallic environments etc.

◆ The logistical conditions, e.g. a large mix of subcontracted suppliers, construction site, character of larger maintenance measures.

◆ The type of assets, e.g. mobile maintenance assets or stationary components; complex or standardized parts, etc., and

◆ The company operations and information flows being supported.

The RFID technologies provide the following functionalities upon which application scenarios can be based:

◆ Identification of maintenance assets or components;

◆ Storage of static or dynamic information on the maintenance assets;

◆ Localization of mobile assets, and

◆ Determination of the condition of maintenance assets over time and through processes.

Benefits of RFID in maintenance: The following examples provide an overview of the benefits attainable by using RFID in maintenance

- ◆ Increased maintenance staff productivity.
- ◆ Reduced manual data acquisition activities and associated error sources.
- ◆ Reduced error sources in process, e.g. mix-ups of similarly designed assets.
- ◆ Elimination of paper documents (material orders, transportation orders, maintenance job orders etc.), printing costs and format changes.
- ◆ Option of integrating third parties, e.g. refurbishing shops, equipment vendors or external maintenance and data exchange with these third parties.

The monetary benefits were determined in the pilot projects by means of cost benefit analysis, however, turnout differently in each case depending on the maintenance process and the company analysis.

Now industries are getting associated with the application of RFID and efforts are aimed at:

- ◆ Establishing the guidelines for the placement of transponder on (initially) selected maintenance components, e.g. engines, transmissions;
- ◆ Specifying reference processes in maintenance that RFID can support;
- ◆ In conjunction with the reference processes, formulating recommendations on the selection of RFID technology;
- ◆ Defining the respective data models and interfaces on Computerized Maintenance Management System (CMMS) for reference process;
- ◆ Addressing other issues such as data security and integrity and process reliability with RFID.

OPTIMIZATION OF MAINTENANCE ACTIVITIES

Here the total maintenance cost is optimized which comprises three items. The cost of maintenance: this is given by the number of times, maintenance is performed multiplied by the cost of each operation. The second term is the cost of failures, this concerns only the direct cost involved in a failure. The last term is the cost of downtime of the system. The sum of the above terms is the overall cost of the maintenance that should be minimized.

Monte Carlo simulation can be performed when the maintenance costs are expressed in terms of variable such as time interval between maintenance, which can be calculated analytically as a function of time and the value that minimizes this cost is selected. At the same time repair resources can also be optimized which will reduce the waiting time of the system/equipment to be maintained.

With the availability of the advanced Monte-Carlo methods, the modelling of complex problems of maintenance, can be undertaken with the interaction of each conflicting parameters.

RISK-BASED MAINTENANCE PLANNING

Risk-based maintenance integrates reliability and risk concepts in the maintenance planning process, thus minimizing consequences and probability of system failure. This methodology is based on quantification of top events resorting to fault trees, a sensitivity analysis of the probability of occurrence of top events and a comparison of corrective measures based on the cost and effectiveness criteria.

The Reliability Centre Maintenance (RCM) is a fairly recent approach aimed at defining maintenance policies based on safety, operational and economic criticality of failures.

However, risk is quantified in coarse classes and the method relies on quantitative or subjective estimates. In fact, simple Failure Model Effect and Criticality Analysis (FMECA) based approaches are usually utilized. Such techniques, however, fully showed their limits in the practical applications.

The Risk-Based Maintenance (RBM), leads to a "Risk-matrix formulation" which, by assigning numerical values to both probability and magnitude parameters, enables to individually rank failure modes from low to high risk thus yielding priority ranking for choosing maintenance tasks. RBM by integrating risk information into the decision-making process enables improved maintenance decisions by focusing resources on highest-risk equipment and providing objective means to evaluate alternative inspection and test strategies thus prioritizing the corrective maintenance activities and enabling reduced costs.

TOTAL PRODUCTIVE MAINTENANCE (TPM)

It is one of the latest mainenance policies adopted in Japan's automotive industry. An adequate maintenance system is planned and implemented, adopting TPM to solve the problems linked to breakdowns microstops, efficiency reduction etc. The TPM method tries to eliminate the main factors of the production losses, the faults, the registrations, the setting changes, the defective pieces. The elimination process of the above causes, starts from the plant reset, then the autonomous maintenance and, finally, programming and planning the preventive maintenance.

TPM is used by the company to modify the preventive maintenance on the basis of the operation results obtained on the field, rather than from the machine producer. The TPM allows structuring the observation, to adopt maintenance forms to train the workers.

Here the maintenance activities are sub-divided into three different levels. The maintenance operations of first level are those of independent maintenance, realized directly by the operator of the machine/system. More complex are the operations of second level, carried out by the maintenance staff and less frequent. The maintenance operations of third level are made sporadically by external staff of machine/system manufacturers.

The first level maintenance includes the verification of the correct operation of the machine/system, to avoid defects in the products. Therefore, not only the "zero-breakdowns" object but the "zero defects" object can also be reached.

The cycle of second level maintenance is constituted by operations that cannot be performed by the machine/system operator but require the maintenance staff intervention. Such operations have to be listed in a specific form of second level maintenance.

The organization of third level maintenance follows such operations which are to be executed by the staff of the machine/system manufacturer with a fixed frequency.

Under this scheme of maintenance the operator/worker is made responsible to carry out the first level maintenance works as he/she remains with the machine/system for longest time, moreover he/she must inform the maintenance staff for remaining operations of second and third level.

MAINTENANCE MANAGEMENT

In the industrial processes, a sudden interruption of the machine impacts on the competitiveness of the company and it is often the most important contribution in the "total production cost". The integrated Decision Support System (DSS) is utilized using predictive maintenance for machine tools to reduce the number of unexpected stops and minimize the life cycle costs of the product avoiding the breakdowns of components. The overall decision support system can have a double impact on maintenance management. In fact it may support:

- ◆ Machine user who can monitor machine performance and ageing of the components; and
- ◆ Maintenance service provider who can plan and forecast interventions needed and optimize maintenance costs.

The system consists of three modules:

1. The diagnostic module that transforms sensor data is useful indicator of the working condition of machine;
2. The ageing module that transforms the previous indicators in an estimate of the wear and the "health state" of the machine;

3. The cost of maintenance module that enables the service provider to interpret the ageing data in order to plan the optimal (technical and economical) maintenance action.

The overall DSS is integrated with Product Data and Knowledge Management (PDKM), which permits to gather and use information from different phases of the machine life. In fact, for example, information about components failures and performance pass from the maintenance provider to designers for new product feature definition. According to information collected directly from previous versions of the product, improvements can be implemented closing the information loop. The other type of information can be transmitted also for recycling and dismissal of product components. The DSS evaluates maintenance costs with an iterative process on each machine for components, exceeding a given threshold value of "criticality". Once the machine user has verified the status of components with the testing module, an alarm on ageing is sent to department which collects the alarms from all the machine tools under maintenance and calculates the economic value of different maintenance actions according to residual life costs estimation. Risk of failure is evaluated and is used to weigh the costs by comparing the actual performance of monitored components with reliability and features supported at the beginning of their life. The system provides value-adding information also to the design department of the machine manufacturer and finally, to the end of the life actors devoted to the reusage, retrieval and dismantling of the components. In fact the DSS calculates the costs of maintenance management not only related to the ageing of the components but relates it to the ageing of the machine.

Further improvements in the modules can provide feedback systems which can be used for improvements, in the maintenance schedules. It can also provide information regarding the availability of each machine tool with the availability of machine provider.

QUALITY CONTROL IN MAINTENANCE

The concept of quality control has been introduced in the area of building construction, i.e. real state management activities with special regards to maintenance services through the achievement of results, which must meet the expectations of quality, effectiveness and efficiency.

The measure of success of maintenance service is a crucial moment because it is difficult to "quantify the quality". That is to define objective parameters and suitable measuring and monitoring instruments capable of quantifying the expected results.

The system followed for the above purpose should be based on the construction of appropriate monitoring indicators and evaluation indexes which will allow to:

◆ Standardize the process of service planning;
◆ Evaluate the outcome of service provided; and
◆ Monitor maintenance costs.

For the above problem two-tier methodologies are adopted.

In the first step the objective is "to measure the quality" and its immediate corollary is "to quantify the quality" by converting it, where possible, into one or more measurable parameters. The second step objective is to "compare the qualities" i.e., the quality required and the quality achieved. Once they have been measured through the introduction of models enabling a correlation between the two sides of evaluation and their own tolerance margins will be possible.

EFFECTIVE MAINTENANCE ORGANIZATION

The success of maintenance function depends on the participation, and total commitment of all employees within the company. The maintenance organization must be established to meet the demands of the operations function. For example, a plant that will be operated 24 hours per day, 7 days per week requires a maintenance organization structure that can support that mode of operation.

Maintenance work force must be distributed to support continuous operation and has an effective planning and scheduling that can effectively take advantage of "windows of opportunity" for example, periods when production demands permit sustaining maintenance activities. On the other hand, when production cycle is 24 hours per day, 5 days per week, the maintenance organization must be configured to take full advantage of the two day window, for example, weekends, to perform sustaining maintenance.

An effective maintenance organization must be organized to provide different levels of maintenance by work type. As a minimum the maintenance organization must be configured to provide effective quality support for three major work type classifications, namely emergency maintenance, preventive maintenance, and periodic rebuilds and overhauls.

To support effective identification, prioritization, planning, and execution of maintenance activities, the organization structure must provide direct and indirect support to the craft work force.

ROOT CAUSE ANALYSIS (RCA)

The concept of RCA is designed to provide cost-effective means to isolate all the factors that directly or indirectly result in the myriad of problems that are faced in the organization. This process is not limited to equipment or system failures, but can be effectively used to resolve any problem that has serious, negative impact on effective management, operation, maintenance, and support of plants and facilities. It has the capability to identify incipient problems, isolate actual cause or forcing function that directly resulted in the problem, as well as identifies all factors that directly or indirectly contributed to the problem.

For example, RCA was used in a steel plant where the cost of the roller bearing went excessively high, which was noticed for a period of six years. The reasons for this high cost were examined through RCA revealed, the primary reason for this was misapplication of predictive maintenance technologies, the results of the analysis indicated that the root causes of the bearing problem were:

♦ Twenty-seven per cent of the bearings in a typical year exhibited abnormal wear because of misapplication.

♦ Improper installation was the cause of twenty-two per cent of the premature bearing failures.

♦ Lubrication related problems contributed eighteen per cent to abnormal failure rate.

♦ Abnormal bearing loading accounted for seventeen per cent of the premature bearing wear.

♦ A variety of other factors, such as electrical arching, caused eleven per cent of the premature bearing wear.

♦ The final five per cent of annual expenditure for replacement of bearing was the result of bearing that had reached the end of their normal life.

Armed with the knowledge gained through the RCA process, corrective actions were taken by crafts to eliminate each of the identified root causes, as well as contributing factors. Crafts were provided training in proper installation and lubrication of rolling element bearings, procedures were rewritten to ensure proper operation and maintenance of the bearings. The specifications and procurement policies were upgraded to ensure proper replacement. Bearings were universally used, and first line supervision was increased to ensure universal adherance to best practices. As a result of these changes, the annual replacement cost of rolling element bearing dropped drastically and remained constant for more than ten years.

Most problems that arise in an organization are addressed in superficial ways, what we call 'first order problem solving'. That is, we work around the problem to accomplish our immediate objective, but do not address the root causes of problem so as to prevent its reoccurrence. By not addressing the root cause, we encounter the same kind of problem again and again, and operational performance does not improve. This is the main reason to use RCA.

The root cause analysis cannot be performed sitting in conference room, office or in front of computer. While the RCA process requires working group meetings, as well as individual and group interviews, the heart of process is gathering factual data that can be used to isolate, identify, and quantify the real reasons that resulted in the abnormal behaviour that is being investigated.

MAINTENANCE OUTSOURCING

Maintenance activities have recently gained a substantial interest due to the strategic management opportunities arisen in the market. The recent developments in the field of maintenance planning and management, demonstrate that the establishment of optimized maintenance policies may drastically improve performance and reduce the operating cost of facilities. However, maintenance activities, typically outside of the core business of production facilities, hence enterprises, often fail to catch the opportunities that may originate by properly optimized management strategies. In such context, market opportunities have arisen for enterprises that make of advanced maintenance management their core business, thus offering their services to companies willing to outsource. A strategic maintenance management may hence encompass the possibility of outsourcing maintenance activities to ensure the necessary availability of production systems, while allowing enterprises to concentrate their sources on their core activities, thus originating economic, financial and operative advantages. As a matter of fact, however, many enterprises are averse to outsourcing strategies: this is frequently due to inadequate or unbalanced contracts.

In order to be effectively undertaken, an outsourcing strategy must be supported by a proper performance-oriented contract. Although much attention has been focused upon the determination of optimal maintenance policies, not much attention has been focused to the importance of a proper establishment of contract variables such as performance level costs, penalties and incentives. While the problem of determining the optimal maintenance policy may be major concern for service provider, the determination of optimal contracting conditions with regard to both the contractors and service provider, may be essential for a successful agreement.

Before the contract is finalized, Service Level Agreements (SLAs) and suitable measureable indicators Key Performance Indicators (KPI) must be decided with associated penalties and incentives/rewards for the work to be performed. A Maintenance Information System (MIS) must also be implemented to allow both the outsourcer and the provider to monitor and control the performance level achieved at each period, thus allowing the evaluation of penalties and rewards.

In order to properly coordinate an outsourcing contract the individual profit functions of the contractors must be calculated and penalties, rewards and performance levels must be properly established in order to ensure the optimal overall performance of the system. To accomplish such objective the possibilities of establishing a win-win strategy between the contractors must be analyzed.

The coordination between the vendor and the outsourcer is possible in the maintenance of global service contracts, provided that penalties and/or incentive related to a performance measure are considered in practice.

QUESTIONS

1. How important is the infrastructure management in the case of maintenance activities? Explain

2. Write down the role of reliability to improve the maintenance functions.

3. How can be the telemetric services applied to maintenance functions? Explain with their merits.

4. Explain how artificial intelligence can help the decision support system.

5. Highlight the use of radio frequency identification used in maintenance functions.

6. What are the benefits obtained by the using RFID in maintenance activities.

7. Using Monte-Carlo simulation how can be the maintenance functions optimized? Explain briefly.

8. What is risk-based maintenance planning and how does it help the maintenance activities?

9. How can the use of total productive maintenance (TPM) help in enriching the maintenance functions?

10. What is the role of management in maintenance function? Briefly explain.

11. Explain how quality control can be exercised in the case of maintenance functions.

12. What is the role of effective maintenance organization in connection with maintenance functions?

13. Explain the Root Cause Analysis (RCA) with suitable example.

14. What do you mean by maintenance outsourcing? Explain it with its limitations.

Bibliography

Agrawal, S.C., *Maintenance Management*, Prabhu Book Service, Gurgaon.

Buffa, L.S., *Modern Production/Operation Management*, Wiley Eastern, New Delhi, 1985.

Collacott, R.A., *Mechanical Fault Diagnosis and Condition Monitoring*, Chapmann and Hall, London, New York, 1982.

Cooling, W.C., *Maintenance Management*, American Management Association, New York, 1974.

Corder, A.A., *Maintenance Management Techniques*, McGraw-Hill, New York, 1976.

Cunnigham, C.E. and Cox, W., *Applied Maintainability Engineering*, A Wiley Interscience Publication, John Wiley and Sons, New York, 1972.

Frank Gradon, *Maintenance Engineering: Organisation and Management*, Applied Science Publishers Ltd., London, 1973.

Frank. L. Evans, Jr., *Maintenance Supervisor's Handbook*, Gulf Publishing Co. Houston, Texas, 1962.

Get, K.H. and Bakh, I.C., *Models of Preventive Maintenance*, North Holland, Amsterdam, 1977.

Gopalakrishnan, P. and Banerji, A.K., *Maintenance and Spare Parts Management*, Prentice-Hall of India, New Delhi, 1991.

Harriss, Ellya, M.J., *Management of Industrial Maintenance*, Butterworths, London, 1978.

Heintzeiman, J.E., *Complete Handbook of Maintenance*, Prentice Hall, Englewood Cliffs (N.J.), 1976.

John W. Criwele, *Planned Maintenance for Productivity and Energy Conservation*, The Fairmont Press Inc., Lilburn, GA 30247, 1987.

Keith, R. Mobley et al., *Maintenance Engineering Handbook*, McGraw-Hill, 2008

Kumar, K., Mishra, R.C. and Ramasubban, P.K., "Reliability Evaluation of Crushing Equipment Used in Coal Processing Plants," *Powder Handling and Processing*, Vol. 2, Number 1, March 1990.

Kumar, K., "Reliability Modelling and Performance Analysis of Power Distribution Systems for Equipment in Coal Mines," Ph.D. thesis, I.S.M. Dhanbad, February 1991.

Mishra, R.C., *Reliability and Maintenance Engineering*, New Age International Publishers, New Delhi, 2005.

Roberto D. Cigolini et al., *Recent Advances in Maintenance and Infrastructure Management*, Springer, 2009.

Shooman, M.L., *Probabilistic Reliability: An Engineering Approach*, McGraw-Hill, New York, 1968.

White, E.N., *Maintenance Planning Control and Documentation*, Gower Press, Teakfield Ltd., England, 1973.

Index